WORLD OF ELECTRONICS

Computers • TV & Video • Radio

Brian Reffin Smith

Christopher Griffin-Beale

John Hawkins

2 What is a computer?

33 TV & video

65 The world of radio

95 Index

This book contains the Usborne Guide to Computers,
the Usborne Guide to TV & Video
and the Usborne Guide to the World of Radio.

Computers was written by Brian Reffin Smith
and illustrated by Craig Warwick, Graham Round, Brian Watson
and Jim Dugdale.
Designers: Graham Round and Iain Ashman
Editor: Lisa Watts

What is a computer?

Put very simply, a computer is a machine which "does things to stuff". In more scientific terms, it is an "information processor". A computer is given information, called "data", instructed to do certain things to it and then show us the results.

The things shown in the pictures below could be called computers. They all receive information which they work on and change to produce new information.

The data put into the computer is called the "input", and the results which come out of it are the "output".

Some people say Stonehenge is a kind of computer. Prehistoric people could work out their calendar from the position of the shadows made by the sun shining on the stones. If you think of the stones as a computer, the sunlight is the input and the calendar is the output.

First modern computer?

This machine might have been the first modern computer – it if had ever worked. It is called the analytical engine and was invented by an English mathematician, Charles Babbage, who lived from 1791 to 1871. Babbage designed the machine to do complicated sums and store the results at each stage in the calculations, and his ideas are the basis for modern computers. The analytical engine never worked, though, because at the time it was not possible to build it accurately enough.

This is a small, modern computer. Information is typed into it via the keyboard and the results are shown on the screen. This computer has worked on information about Stonehenge to produce a picture.

Your brain is a computer in that it receives information from the eyes, ears and other senses (smell, touch and taste) and sends out instructions for action.

DISPLAY SCREEN

THE COMPUTER'S WORK IS DONE IN HERE

KEYBOARD INPUT

Electronic chess games have a very small computer inside. From the position of the pieces at any stage in the game, they can work out the best move.

This digital watch can compute what time it is anywhere in the world. Given the time in London, say, it can work out what time it is in New York.

Input and output

There are lots of different ways of putting information into a computer, and of getting the results from it. The small computer on the opposite page has a built-in keyboard for input and screen for output.

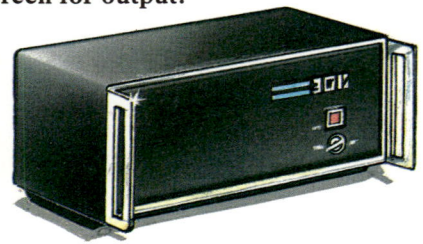

This black box is a computer which can have many different kinds of input and output equipment plugged into it. Here are some examples.

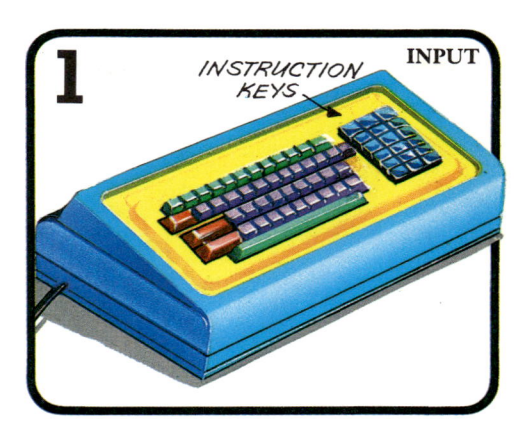

1 INPUT

A keyboard input has letters and numbers like a typewriter, and also a set of instruction keys. It may be connected to a screen or printer (see below) so you can see what you are typing.

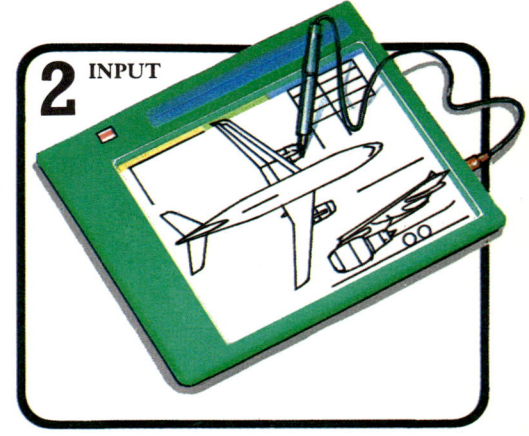

2 INPUT

Information, such as technical drawings and graphs, can be put into a computer by drawing on a magnetically sensitive surface with a special pen. This is called a graphics pad.

3 INPUT

Look out for bar codes, like this, on packages in stores. Details about the product are given to a computer by scanning the lines and spaces with a low power laser beam from a special pen.

4 INPUT

You can also instruct computers, or give them information, by talking into a microphone. At present they can cope with only a limited number of words in a voice they are trained to recognize.

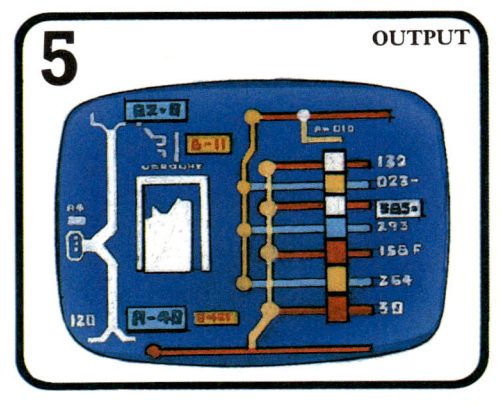

5 OUTPUT

Complicated information coming out of a computer can be shown as graphs, diagrams and words on a screen like a television screen. This is called a visual display unit, or VDU.

6 OUTPUT

Printers, like this one, print the computer's output on paper. When linked to a powerful computer they have to be very fast to keep up with the flow of information and some can print 2,000 lines a minute.

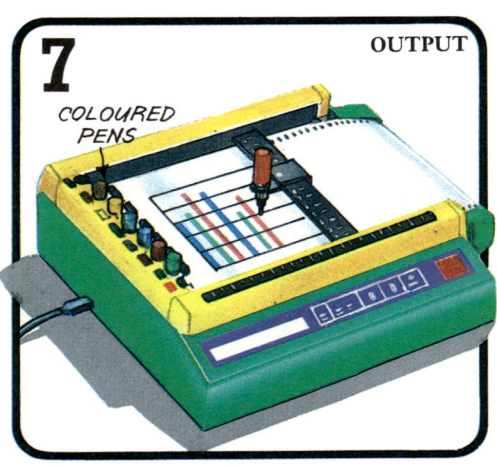

7 OUTPUT

Graphs, pictures or words can be drawn on a "plotter" by signals from the computer guiding the pen across the paper. Some plotters can automatically pick up a pen of a new colour.

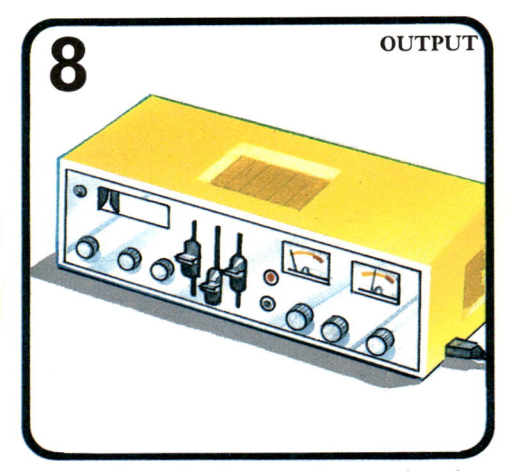

8 OUTPUT

A sound synthesizer, acting on signals from the computer, can put sounds together to make words. It is easier for a computer to "talk" than for it to recognize spoken words.

Types of computers

Once, all computers were big, expensive and used a lot of power. They were called "mainframes" because the parts were mounted on frames in large metal cabinets. The large, powerful computers of today are still called mainframes, but now there are also smaller machines called "minis" and even smaller, desk-top ones called "micros". Over the last 40 years computers have become smaller and smaller and more and more powerful.

Whatever their size, all computers have the same basic parts. These are described in the picture on the opposite page.

1 Mainframes

The equipment for a large modern computer can fill several rooms. There are rows of data storage cabinets containing information for the computer, and many different kinds of input and output equipment, such as printers, VDUs and keyboards. A modern mainframe can carry out millions of instructions every second, and works so fast that it can do many different jobs at once.

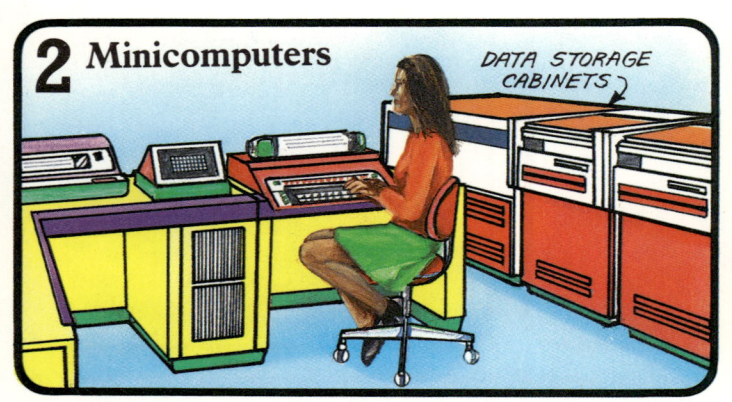

2 Minicomputers

A minicomputer is smaller than a mainframe and cannot handle so much data, or work as fast as the giant computers. A modern minicomputer, though, is many times more powerful than the vast mainframes of the early days of computers. Minicomputers are generally used for one particular type of work, whereas a mainframe does lots of different jobs.

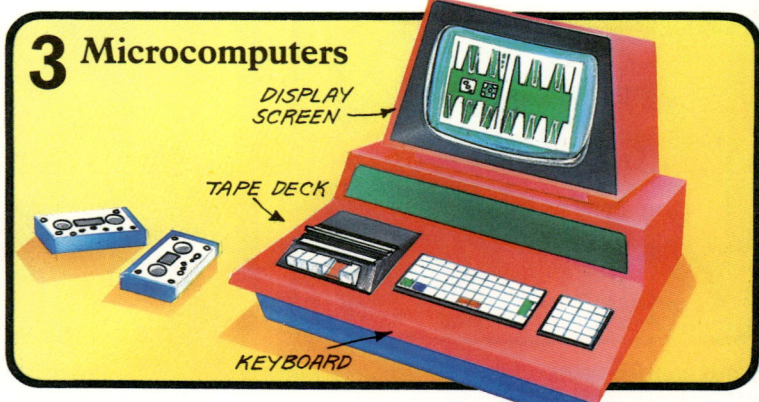

3 Microcomputers

When smaller, cheaper microcomputers, like this, were developed, many more people could afford a computer. Present-day micros are not as powerful as the larger computers, but most can be connected to extra equipment so they can store more information, and to other kinds of input and output machinery, such as plotters and printers.

First electronic computers

Today, all computers are electronic, that is, all their work is done by pulses of electricity. Work on the first electronic machines began in the 1940s, in an attempt to create computers which could crack enemy codes and work out target distances for the artillery in World War II.

One of the first American electronic machines, called ENIAC, was built in 1946. It could do five thousand calculations a second, but it was not a true computer as it could not store information or instructions. It was thousands of times faster, though, than a mechanical calculator, that is, one which works by means of moving gears and wheels.

This is the Manchester University Mark I, an early British electronic computer. It was not as fast as ENIAC (it could do 800 calculations per second) but it could store the instructions for carrying out a series of calculations and because of this it is considered to be the first true computer. It was built with electronic parts left over from World War II and first ran on 21 June, 1948, for 52 minutes.

Parts of a computer

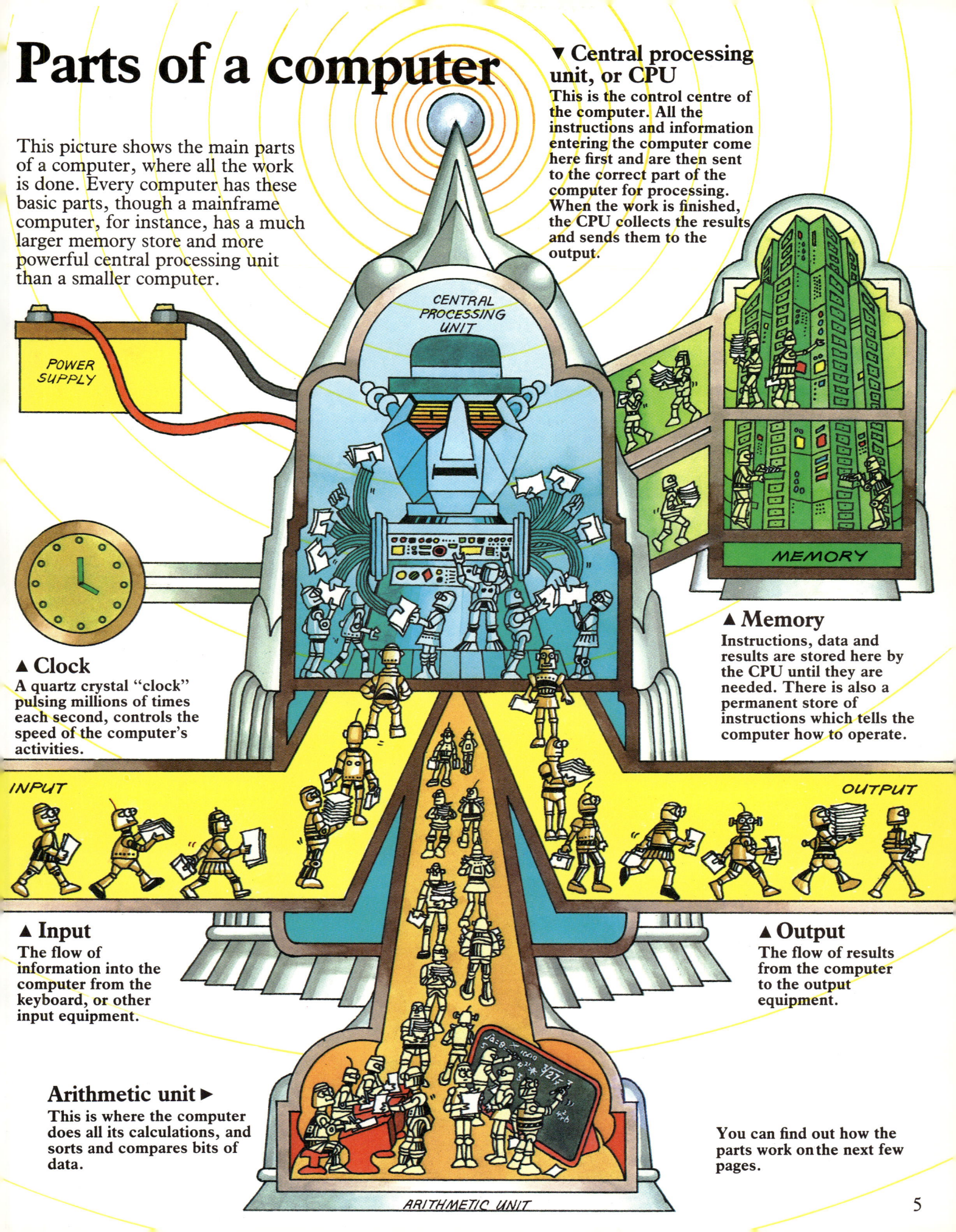

This picture shows the main parts of a computer, where all the work is done. Every computer has these basic parts, though a mainframe computer, for instance, has a much larger memory store and more powerful central processing unit than a smaller computer.

▼ Central processing unit, or CPU
This is the control centre of the computer. All the instructions and information entering the computer come here first and are then sent to the correct part of the computer for processing. When the work is finished, the CPU collects the results and sends them to the output.

▲ Clock
A quartz crystal "clock" pulsing millions of times each second, controls the speed of the computer's activities.

▲ Memory
Instructions, data and results are stored here by the CPU until they are needed. There is also a permanent store of instructions which tells the computer how to operate.

▲ Input
The flow of information into the computer from the keyboard, or other input equipment.

▲ Output
The flow of results from the computer to the output equipment.

Arithmetic unit ▶
This is where the computer does all its calculations, and sorts and compares bits of data.

You can find out how the parts work on the next few pages.

Inside a computer

The pulses of electricity which do all the work inside a computer are controlled by parts called electronic components. The components in the first electronic computers were called valves. In the 1950s, a new kind of component, called a transistor, was invented. With transistors it was possible to build smaller, much more reliable computers. The greatest advances, though, came with the invention of integrated circuits, or "chips" as they are often called. An integrated circuit is a tiny chip of a substance called silicon on which there are thousands of components packed closely together.

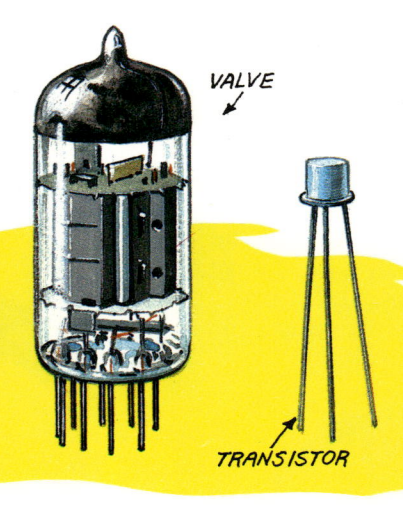

Valves, transistors and chips

Valves, like the one shown above, used a lot of power and got very hot. Transistors used much less power than valves, so they did not get so hot and could be packed closer together in smaller computers. An integrated circuit, shown here in the plastic case which makes it easier to handle, contains thousands of minute transistors linked together to make paths, called circuits, through which the electric current passes.

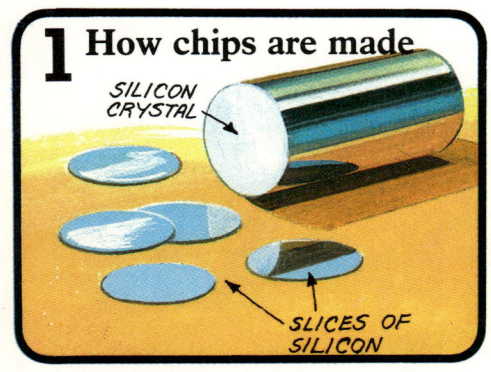

1 How chips are made

To make the chips, crystals of silicon 99.9999999% pure are grown in a vacuum oven. The silicon is so pure that it will not conduct electricity until it is treated with certain chemicals. The silicon is cut into slices and up to 500 chips will be made from each slice.

2

The circuit containing the components for a chip is designed with the help of a computer. It is drawn out 250 times larger than it will be on the chip. Some chips have eleven or more different circuits containing tens of thousands of electronic components, built up one on top of the other in the silicon.

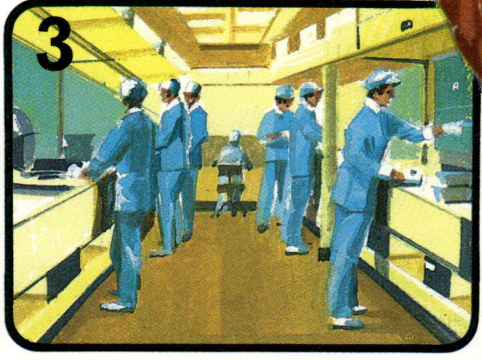

3

Then the circuit design is reduced to chip size and photographically copied lots of times on to each slice of silicon. This is done in ultra-clean, air-conditioned laboratories which are about a hundred times cleaner than a modern hospital operating theatre, so no dust gets on the circuits.

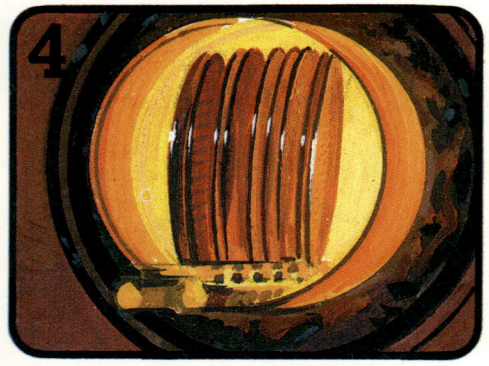

4

The silicon slices are placed in a furnace at a temperature of over 1000°C and exposed to certain chemical elements. In the great heat of the furnace, atoms of the chemicals enter the surface of the silicon, but only along the lines of the circuits.

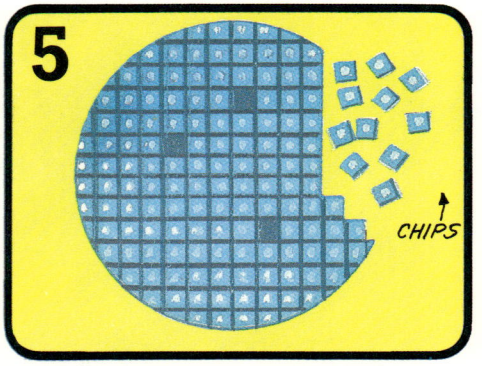

5

Stages three and four are repeated several times until each chip contains several different circuits of chemically treated silicon through which electric current can pass. The circuits are tested – up to 70% are marked as faulty – and then the slices are cut up into chips with a diamond or laser saw.

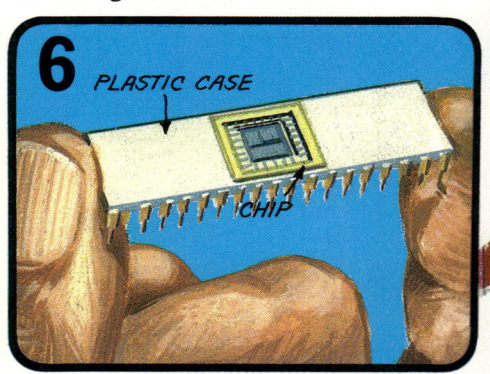

6

Each tiny chip is then put in a plastic case with gold wires connecting the circuits in the silicon to the pins on the case. This makes the chip easier to handle and fit into the equipment it will eventually be part of.

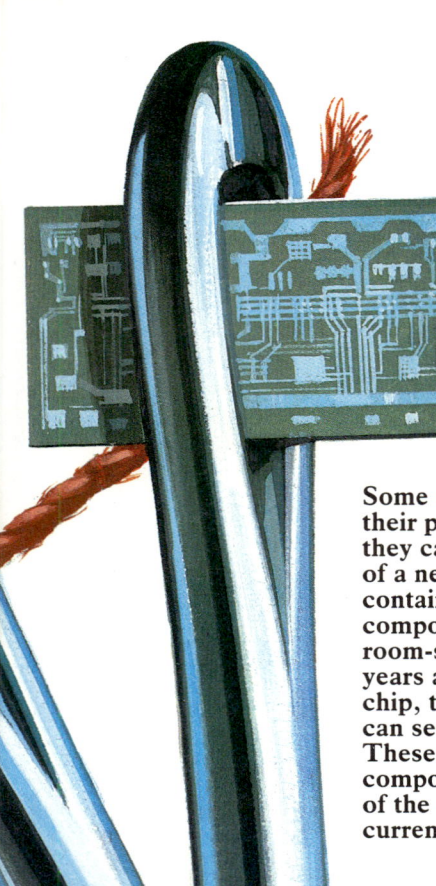

INTEGRATED CIRCUIT ON A CHIP OF SILICON

Some silicon chips (without their plastic cases) are so tiny they can fit through the eye of a needle, yet each chip contains more electronic components than the room-sized computers of 30 years ago. On this enlarged chip, the pattern of lines you can see are its circuits. These contain the components and are the part of the chip through which the current passes.

Kinds of chips

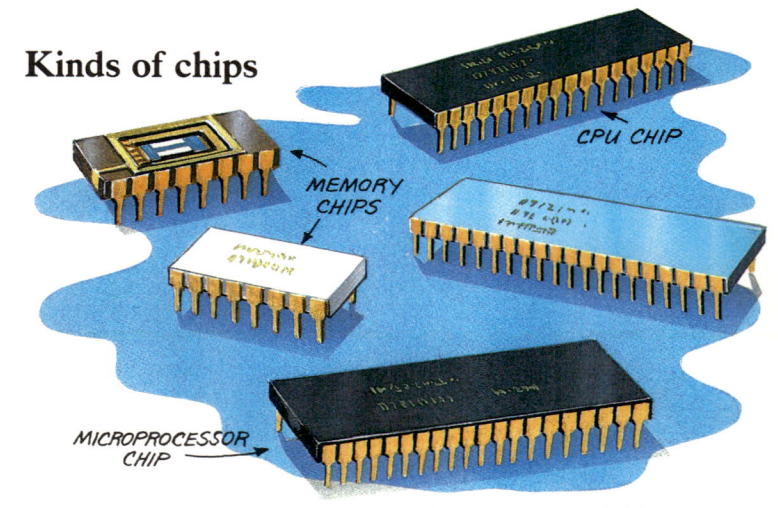

CPU CHIP
MEMORY CHIPS
MICROPROCESSOR CHIP

There are lots of different kinds of chips and each kind has circuits specially designed to do a certain job. There are special chips for the central processing unit of the computer and for the memory store, and others to do the work in the arithmetic unit. Some chips have circuits which can do the work of all the different parts of a computer. They are called microprocessors.

Computer on a chip

This is an enlarged picture of the circuits on a microprocessor chip, showing the parts which do the same work as a computer. Chips like this are used in computers, as well as in calculators, electronic games, and equipment such as washing machines.

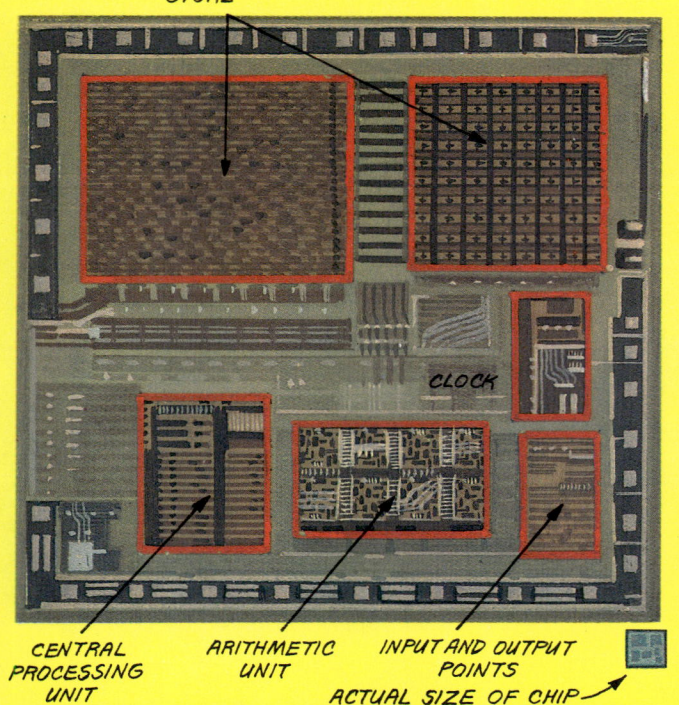

MEMORY STORE
CLOCK
CENTRAL PROCESSING UNIT
ARITHMETIC UNIT
INPUT AND OUTPUT POINTS
ACTUAL SIZE OF CHIP

1 Building a computer

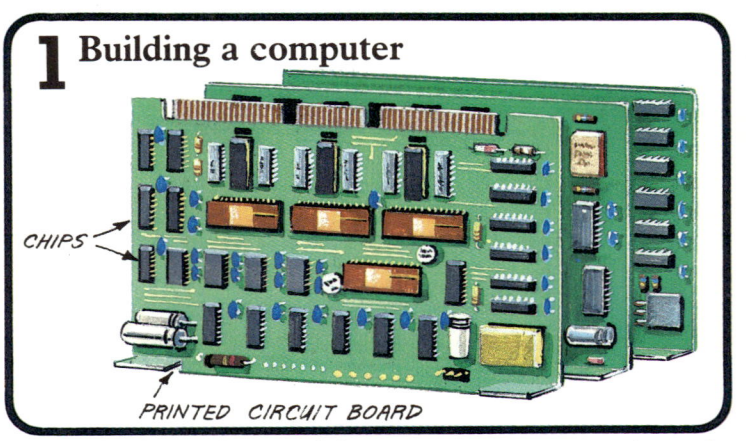

CHIPS
PRINTED CIRCUIT BOARD

The chips for each part of the computer are mounted together on boards called printed circuit boards. The chips are connected by narrow bands of metal printed on the board, which carry the electricity to the chips. Then the boards are put together to make the computer.

2

POWER SUPPLY
PRINTED CIRCUIT BOARDS
MEMORY CHIPS

This photograph shows the inside of a small computer. You can see how little space the circuit boards need. Silicon chips, which are very cheap to produce, have made it possible to build small, quite powerful computers like this more cheaply than ever before.

How computers work

How can a computer which contains only a mass of silicon chips, process a number, a word, or even a picture? The answer is that the electric current passes through the chips in series of pulses which form a code which can represent anything at all – numbers, letters or even colours. The code of pulses is created by the transistors in the chips. The transistors act like switches, turning the current off and on. While the computer is working, millions of pulses are passing through the circuits of the chips every second.

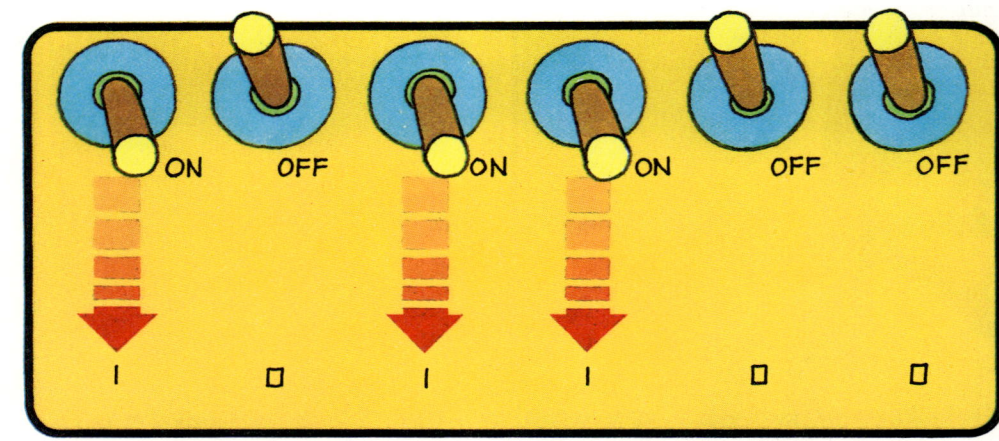

There are only two signals in the code used by computers: pulse and no-pulse, or "on" and "off". It is called binary code and another way of expressing it, which makes it easier to write down, is with the digits "1" for pulse and "0" for no-pulse.

Counting in binary code

The numbers we normally use are called decimal numbers, but you could write numbers in binary code instead.

In the decimal system there are ten digits and the system is based on tens. That is, each of the digits in a number is ten times the value of the digit on its right. For instance, the number 1,463 means, reading from the right:

In the binary system there are two digits and the system is based on twos. Each of the digits in a number is twice the value of the digit on its right. For instance, the binary number 1111 means, reading from right to left:

3 lots of 1	=	3
6 lots of 10	=	60
4 lots of 100	=	400
1 lot of 1000	=	1000
which added together	=	1463

or one thousand, four hundred and sixty-three.

1 lot of 1	=	1
1 lot of 2	=	2
1 lot of 4	=	4
1 lot of 8	=	8
which added together	=	15

So 1111 in binary is fifteen in our number system.

Finger computer

Here is an easy way to change binary numbers to our numbers.

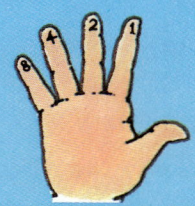

Hold your right hand up with the palm towards you. With a felt-tip pen, write "1" on your first finger, "2" on the second finger, "4" on the third and "8" on the fourth.

To use your finger computer, stick fingers up for binary "1s" and fold fingers down for binary "0s". Then add the numbers on the fingers sticking up and the total is the answer in decimal numbers.

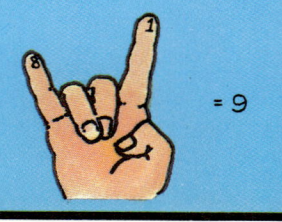

How computers use the code

The stream of pulses travelling through the circuits is controlled by the transistors switching on and off, sending pulses on round the circuit, or holding them back. These transistor switches are also called gates, and there are lots of different kinds. A simple gate has only two points, called terminals, where it receives pulses. Whether or not it sends on a pulse depends on the pulses it receives.

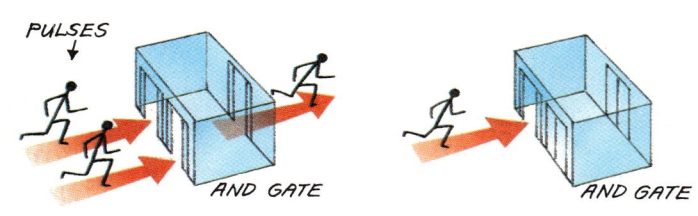

One kind of gate sends on a pulse only when it receives a pulse at both its terminals. This is called an AND gate.

Another kind, called an OR gate, sends on a pulse when it receives one at either or both of its terminals. A NOR gate only sends one on if neither of its terminals receives a pulse.

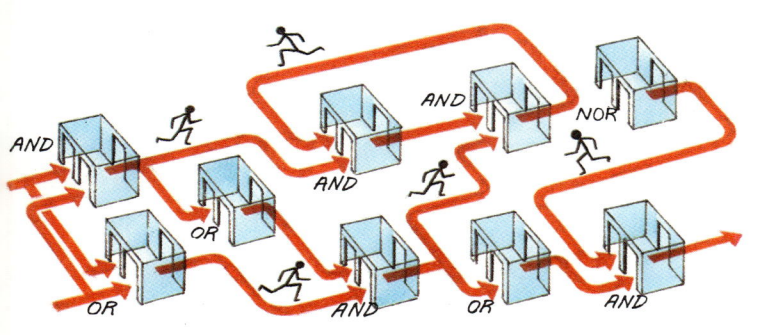

Thousands of these gates are arranged in circuits to create patterns of pulses which can add, compare, memorize and do all the other work inside the computer.

Computerized space pictures

With a code of hundreds of thousands of pulses, the computer can deal with almost anything. For example, given a fuzzy, indistinct picture of a planet taken from a satellite, the computer can work out the details in the picture.

The computer is given a picture like this and told to analyze all the different shades in the picture. Then it is instructed to make all the areas of the picture in shade one, red, all the areas in shade two, orange, and so on.

By repeating this process lots of times the computer can make a picture, like the one on the right, which shows the planet more clearly. The colours are used to make the shape clearer and are not the actual colours of the planet.

Computer doodles

Here is a new way to make doodles. Draw a set of squares on squared paper, as shown below. Then, using the rules shown on the right, draw another version of the doodle and see how it changes. This is how a computer changes pictures – by altering each little dot in the picture according to the rules it is given.

1. CHANGE BLACK SQUARES WHICH DO NOT HAVE EITHER TWO OR THREE BLACK NEIGHBOURS TO WHITE.

2. CHANGE WHITE SQUARES WHICH HAVE THREE BLACK NEIGHBOURS TO BLACK.

Remember, each square has eight neighbours – one above, one below, two sideways and four diagonally.

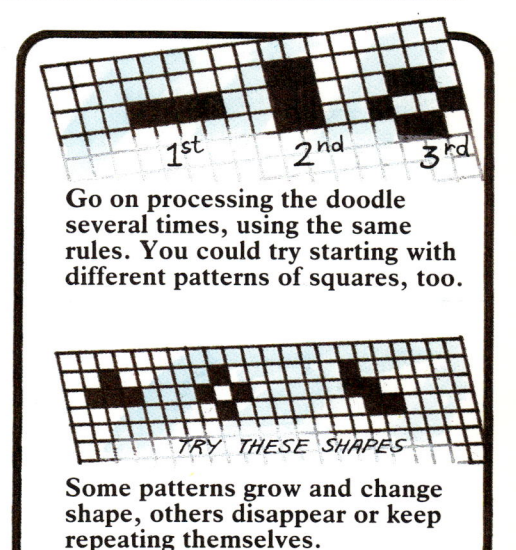

Go on processing the doodle several times, using the same rules. You could try starting with different patterns of squares, too.

TRY THESE SHAPES

Some patterns grow and change shape, others disappear or keep repeating themselves.

The computer's memory

In the electronic circuits of its memory, the computer holds a vital store of instructions, data and results. This ability to store information enables it to carry out very difficult calculations by working through them step-by-step, storing the results of each stage and checking and comparing them with later results and information.

The computer's built-in memory is not very large, but it can also store information on discs and tapes, which can be put into the computer when needed. These are called the computer's backing store.

Information, both inside and outside the computer, is stored in binary code. The binary digits "1" and "0" are also called "bits". To record all the letters of the alphabet, as well as numbers and signs, the computer needs more than the four combinations: 10, 01, 11, 00, which are possible with the two bits. So each letter or number is usually represented by a group of eight bits, called a "byte". The memory of a small computer can store about 64,000 bytes (that is about 12,000 words, or half a novel), and a large computer can store millions.

Built-in memory

Inside the computer there are two types of memory. One, called ROM, is a permanent store of instructions telling the computer how to work. The letters stand for Read Only Memory. The computer can only read the information in ROM, and you cannot usually rub it out or put new information there. The instructions in ROM are built into the computer when it is made.

The other type of memory is called RAM, which stands for Random Access Memory. This is where the computer stores all the data and instructions it receives from the input, and the results as it works through its calculations. RAM is a short-term memory – when the computer is switched off, all the information disappears, but the ROM memory stays intact.

Human memory

Like computers, people also probably have a permanent, or long-term memory, and a short-term memory (called LTM and STM for short). Here is a memory test to try out on a friend.

On two large pieces of paper, write the letters as shown below.

TREPTOSESLANTBI
KLERNEFOPLUFTO

1 Say to your friend: "I am going to show you some letters for a few seconds. Wait until I give you the signal, then write down what you remember." Hold up one set of letters for about five seconds. Cover them, wait ten seconds, then signal by tapping something.

2 Now repeat the test with the other letters. This time, instead of giving a signal, say: "OK, you can write them down now". The person probably gets fewer right this time, as your spoken instruction had to be stored in STM as well, and it pushed out some of the letters.

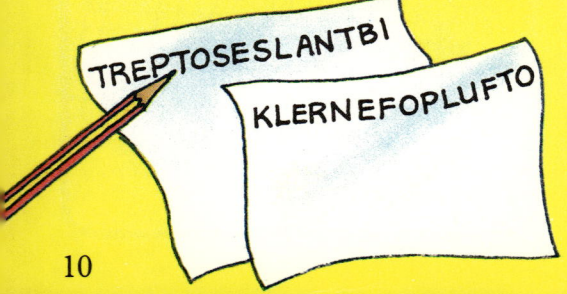

Memory store outside the computer

No computer yet built has a large enough memory to hold all the information it needs, and all the details in the memory store are lost when the computer is switched off. An unlimited amount of information, though, can be held permanently in the computer's backing store.

The backing store can consist of magnetic discs, or tapes, or punched cards or tapes on which the computer can record and read back information.

A very large backing store of information for the computer is called a databank. With new inventions, such as the bubble memories and laser discs shown below, millions of words can be stored in a very small space. If all the information in a databank was stored in books, it would fill two or three large libraries.

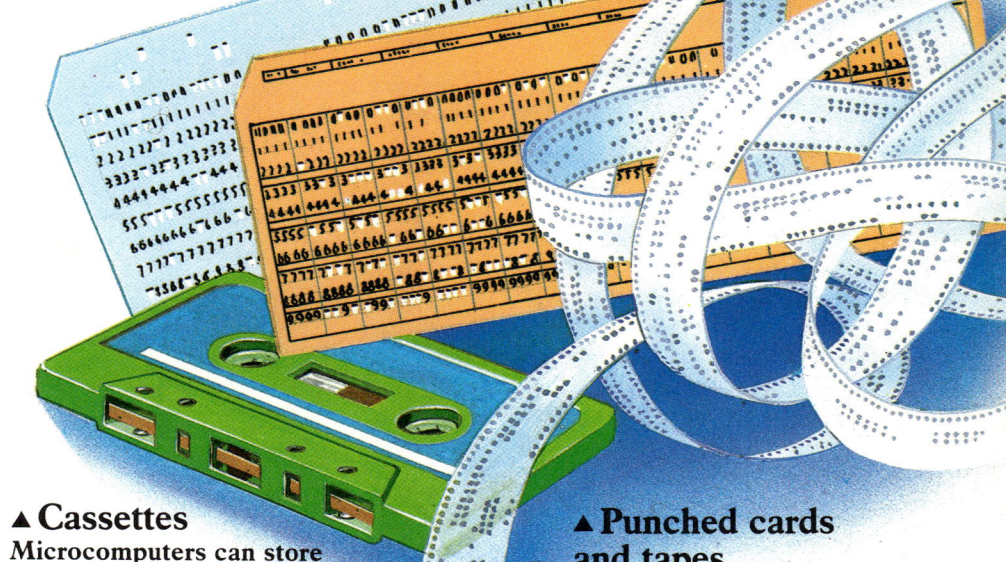

▲ Cassettes

Microcomputers can store information on the magnetic tape in a cassette, using an ordinary cassette recorder. The information is stored in binary code with a high sound for a "1" and low sound for "0".

▲ Punched cards and tapes

With these, information is stored as patterns of holes which the computer can "read". Each hole represents a binary "1".

Bubble memories

These are special chips which store information as tiny "bubbles" of magnetism. Strings of "bubbles" and "no-bubbles" represent the information in binary code, and each chip can store hundreds of thousands of bubbles.

▲ Floppy discs

These are plastic discs with magnetically sensitive surfaces. The computer can pick bits of data from anywhere on the disc, whereas a tape has to be run from the beginning.

HOLDS ABOUT 256,000 BYTES

Laser discs

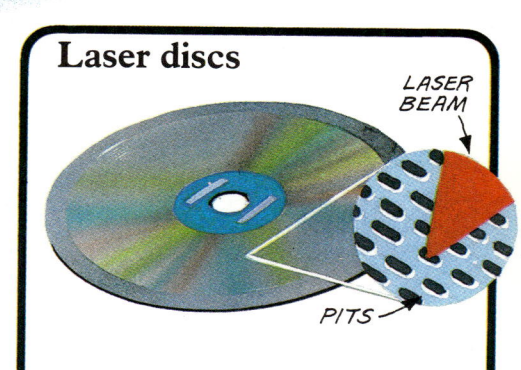

On these, information in the form of binary code is stored as microscopic pits which can be "read" by a laser beam. Each disc can hold 80 million words (twice as many as in the whole of the *Encyclopaedia Britannica*).

How to write messages in binary code

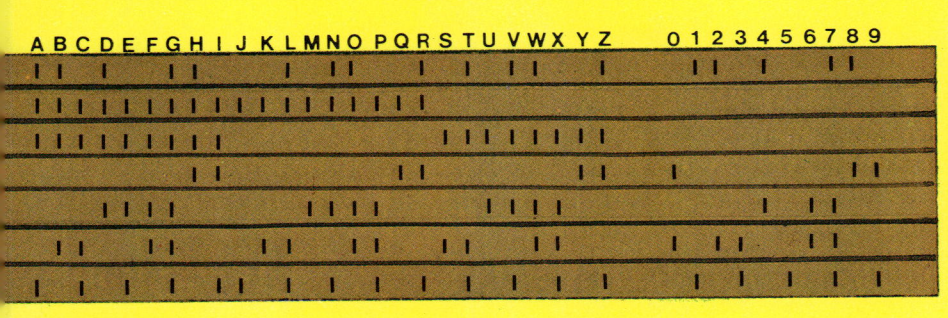

DRAW SIX LINES

This diagram shows the pattern of pulses (dark dashes) representing the letters and numbers on a piece of magnetic tape. Each letter and number is shown by a vertical column of pulses and no-pulses. The pulses are binary "1s" and the no-pulses are "0s".

To write messages in this code, draw lines across a piece of paper, as shown. Then, for each letter in the message, make dots between the correct lines as shown in the guide on the left. Leave spaces between each word. Can you work out the word shown above? The answer is on page 32.

Telling the computer what to do

A list of instructions telling the computer what to do is called a program.* Some programs control the basic operations of the computer, and these are stored away permanently in its ROM memory. Others, telling it exactly what to do for a specific job, have to be specially written. Each program has to be very carefully worked out as any errors will lead to mistakes in the computer's work.

Programs, and all the data fed into the computer are called software, whereas the computer equipment is called hardware.

1 Stupid peanut program

NO PEANUTS

1. LEAVE HOME.
2. GO TO SHOP, ASK FOR PEANUTS.
3. IF SHOP HAS NONE, GO BACK TO LINE 2.
4. GO HOME.

Here is a list of instructions for buying peanuts, written as though it were a program for a computer. There are some mistakes in this program, though. Can you spot them? Mistakes in a program are called "bugs".

2

There are two bugs in the peanut program. Line 2 does not tell the computer to try a new shop, so it would probably keep going to the same shop and asking for peanuts.

3

The other bug is in line 3. It does not tell the computer when to stop, so even if there were no peanuts anywhere, the computer would still go on trying.

4

1. LEAVE HOME
2. GO TO NEAREST SHOP NOT ALREADY VISITED. ASK FOR PEANUTS.
3. IF THE SHOP HAS SOLD OUT AND IF YOU ARE NOT TIRED, GO TO LINE 2.
4. GO HOME.

This is a better version. At line 3, if the shop has peanuts, or if the computer is tired, it does not go back to line 2. Instead it goes to line 4 which sends it home.

How to make a computer which writes funny poems

Here, and on the next two pages you can find out how to make a cardboard "computer" which can write 16,384 different poems.

You will need a strip of paper 60cm x 6cm (tape several pieces together to make a strip this long), a piece of thin card 12cm x 20cm, more paper to write on and a pencil, rubber and scissors. You will also need another small piece of card and a used matchstick.

This page shows you how to make the computer and write the program, then, over the page, you can find out how to use the program to write funny poems.

1

Cut two slits in a piece of thin card, as shown in the picture. The slits should be about 5cm apart and should measure about 7cm across.

2

Thread the strip of paper through the slits like this and pull it down so 5cm sticks out the back at the top. If it does not slide easily, make the slits wider.

*Never "programme" in computer jargon.

Programming languages

You could program a computer directly in binary code, but it would be very difficult to do. Instead, programs are written in special programming languages which the computer can understand. It has a master program built into its memory which translates the programming language.

Many different programming languages have been invented to suit different sorts of problems. For instance, FORTRAN is for mathematical and science-based problems, COBOL is for business work and POP2 is good for logical problems. Here are some other languages.

```
400 PRINT "ENTER CO-ORDINATES"
410 N=0
420 INPUT X(N), Y(N)
430 IF X(N)=0 AND Y(N)=0
440 N=N+1:GOTO 420
450 FOR I=1 TO N
460 X(I)=X(I)+100
470 NEXT I
480 PRINT"ENTER ROTATION
490 INPUT RX,RY,RZ
500 PRINT"PLOTTER <P>
510 INPUT Z$
```

This is part of a program, written in a language called BASIC, telling the computer how to make the picture on the right. BASIC is good for lots of different kinds of problems, and it is quite easy to learn. A lot of the terms are based on English words and mathematical symbols. Each of the instructions in a program is usually numbered. To begin with, the numbers go up in tens so more instructions can be added between them if necessary.

PILOT is a good language for writing the programs to teach subjects in schools. Programmed in PILOT, the computer can recognize answers expressed in many different ways from different students.

EXPLOR was invented to help artists write programs telling the computer how to gradually change the shapes in a pattern to make a new design.

LISP is a special language used to program machines to try and make them do certain things. This robot can be programmed to search for a power socket with its TV camera "eye" and plug itself in to recharge its batteries.

3
Now you are ready to write the program instructions on the strip of paper. First, write instruction one, as shown above, on the paper showing between the slits.

1. A = 0
   B = 0

4
Pull the paper up so instruction one disappears, and write instruction two, given above. Leave only about 2cm between the instructions.

2. ADD 1 TO A

5
3. IF A=6 GO TO LINE 10
4. WRITE DATA LINE A
5. ADD 1 TO B
6. TWIRL SPINNER TO FIND N
7. WRITE DATA WORDS FROM ROW B COLUMN N
8. IF B=3 OR 5, GO TO LINE 5
9. GO TO LINE 2
10. STOP

Continue pulling the paper up and writing all the instructions given above, on the paper strip. These instructions are explained on the next page.

Continued on the next page ▶

Writing a computer program

Writing the program is one of the most important stages in solving a problem with a computer. First you have to study the problem very carefully and work out what information the computer will need, and what steps it must go through to solve the problem. Sometimes you also draw a chart, called a flowchart, showing the sequence of the steps in the program. The computer can act on only one bit of information, or one instruction, at a time, so you have to work the program out very precisely and make sure the steps are in the right order.

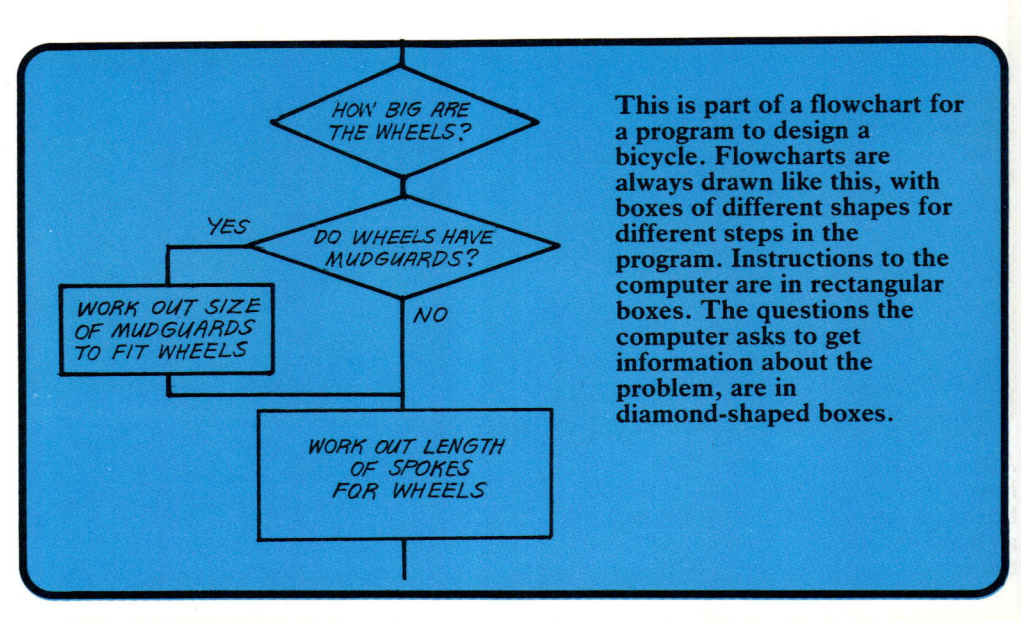

This is part of a flowchart for a program to design a bicycle. Flowcharts are always drawn like this, with boxes of different shapes for different steps in the program. Instructions to the computer are in rectangular boxes. The questions the computer asks to get information about the problem, are in diamond-shaped boxes.

▶ **Funny poem computer continued**

The program for the computer should look like this now, with all ten instruction lines written down the strip of paper.

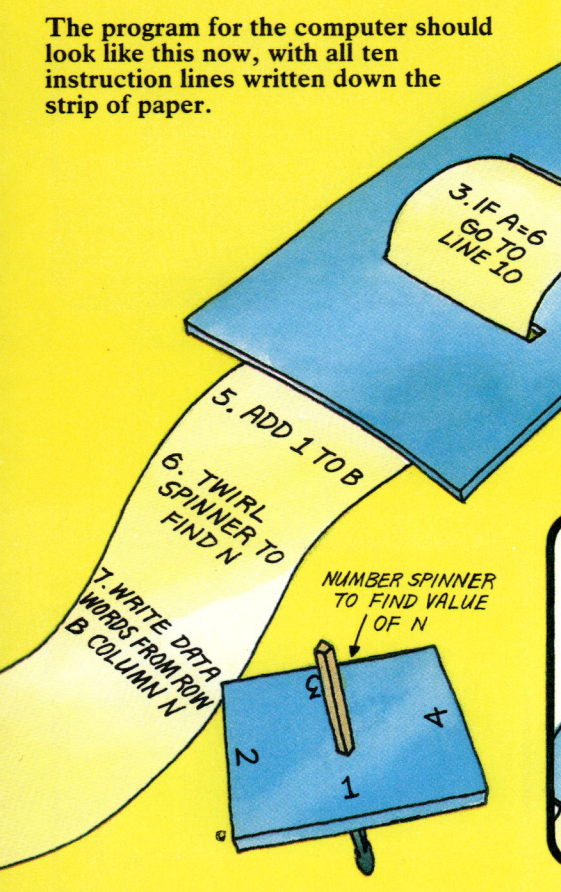

How it works

The letters A, B and N represent numbers. At the start of the program A and B are zero, but as you work through, the instructions tell you to add one to A and B, so the numbers change. The numbers show you which data lines and words to use from the lists on the opposite page.

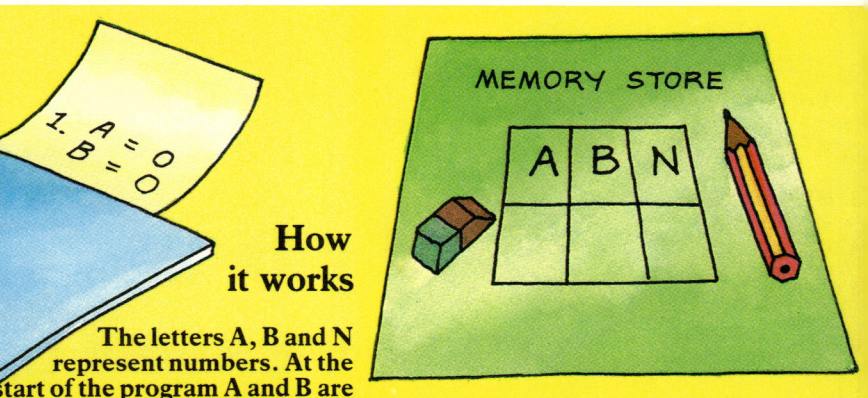

To remember the values of A, B and N, draw a "memory store" chart. Pencil the numbers in and rub them out and change them as you work through the program.

To make the spinner, write numbers like this on a small square of card and poke a used matchstick through the centre. When you twirl the spinner, the side it leans on when it stops gives you the number for N.

1 Running the program

Set the program to line one and, as instructed, write zero in the memory store for A and B. Now work down through the program and whenever it tells you to add one to A or to B, change the number in the memory store.

2

When the program tells you to "Write data line A", look in the memory store to find the number for A. Then, on another piece of paper, write out the data line which is the same number as A.

14

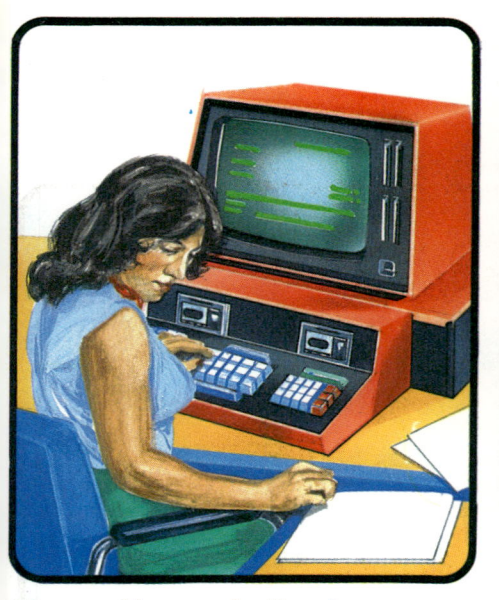

```
RUN .............................................(tells computer to run through
                                                  program)

SYNTAX ERROR IN LINE 20........(computer replies that there is a
                                                  mistake in line 20)

LIST 20 ........................................(operator asks to see line 20)

20 PRONT "HOW MANY?" ..........(computer displays line 20 and shows
                                                  that word "print" is wrong)

OK................................................(computer says it is ready for next
                                                  instruction)

20 PRINT "HOW MANY?"............(operator types in new line)

RUN .............................................(tells computer to run through
                                                  program again)
```

After working out the flowchart, you have to translate the contents of each box into a programming language such as BASIC. You type the program into the computer via the keyboard.

Then you tell the computer to run through the program and display it on the screen. Very few programs are correct first time, and they have to be "dubugged" to remove the mistakes. These may be errors of logic in the steps in the program, or typing mistakes, as in the part of the program shown above. Once the program is correct, you tell the computer to run through it again, and as the questions in the program appear on the screen, you type in the information the computer needs.

Data lines

1. THERE WAS A YOUNG MAN FROM
2. WHO
3. HIS
4. ONE NIGHT AFTER DARK
5. AND HE NEVER WORKED OUT

These are the "data lines" for the poem. When the program tells you to "Write data line A", find the value of A from the memory store, then write out the data line of the same number.

Data words

	Column 1	Column 2	Column 3	Column 4
1	TASHKENT	TRENT	KENT	GHENT
2	WRAPPED UP	COVERED	PAINTED	FASTENED
3	HEAD	HAND	DOG	FOOT
4	IN A TENT	WITH CEMENT	WITH SOME SCENT	THAT WAS BENT
5	IT RAN OFF	IT GLOWED	IT BLEW UP	IT TURNED BLUE
6	IN THE PARK	LIKE A QUARK	FOR A LARK	WITH A BARK
7	WHERE IT WENT	ITS INTENT	WHY IT WENT	WHAT IT MEANT

These are the data words to complete each line of the poem. Each row contains words suitable for one of the lines. The words you use are decided by the values of B and N as you work through the program. B gives the number of the row and N is the number of the column.

3 When it says "Write data words row B, column N", find the numbers for B and N from the memory store. Then complete the line of the poem with the data words from row number B, column N.

4 For lines three or eight, if your number for A or B is not the same as the numbers given in the program, miss out that instruction and move on to the next line. This is just how a real program works.

5 Work backwards and forwards through the program until you get to line ten and the poem is finished. Then, if you work through the program again, the "computer" will give a different version of the poem.

Things computers can do

The great speed with which computers can work through vast amounts of information makes them good for calculating millions of telephone bills, keeping business records of sales and payments, for scientific calculations, and so on. Here, though, are some other things computers can do.

Teaching in schools

These children are using a computer to help them learn to read and write. The children answer the computer's questions by touching a sensitive board near the word or picture they think is correct.

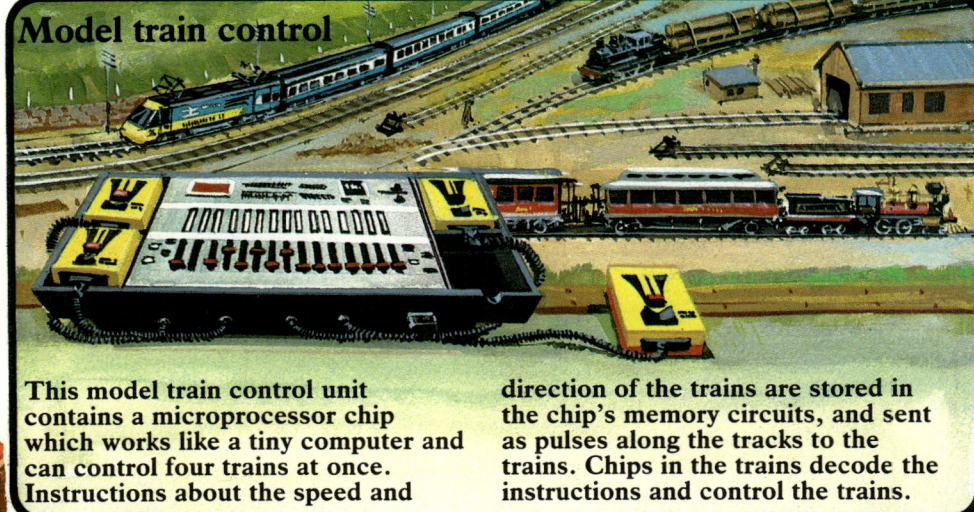

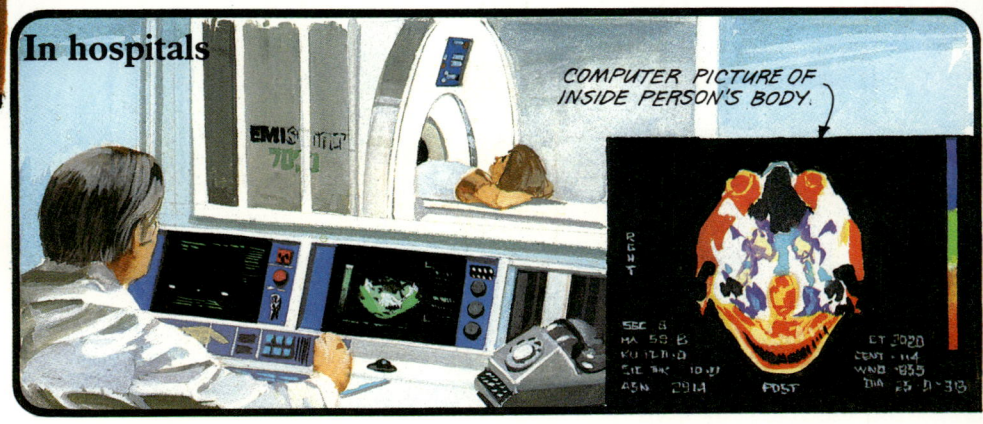

Model train control

This model train control unit contains a microprocessor chip which works like a tiny computer and can control four trains at once. Instructions about the speed and direction of the trains are stored in the chip's memory circuits, and sent as pulses along the tracks to the trains. Chips in the trains decode the instructions and control the trains.

In hospitals

COMPUTER PICTURE OF INSIDE PERSON'S BODY.

A computer can show detailed pictures of the inside of a patient's body, based on photographs taken by an X-ray scanner. The scanner takes thousands of pictures of the patient, from lots of different angles. The pictures are sorted and processed by the computer, and the doctor can then ask it to show an organ in the body from any angle.

Weather forecasting

Computers have only recently become powerful enough to help produce weather forecasts. Weather stations and satellites all round the world send in frequent reports of the changes in the winds and temperature which affect our weather. The computer has to analyze all this data and continually adjust its predictions as the conditions change.

Helping the handicapped

Severely handicapped people can use computers to communicate with other people, "talking" to them via the computer screen. In the system shown above, the person controls the computer by sucking and blowing down a tube, telling it which letters to select to spell out words.

Disaster relief

Computers help organize the sending of aid to the victims of earthquakes or famines. The computer can sort through all the information about the disaster area, and record what goods have been sent. It can then show, for instance, that more food rather than blankets is needed, and that several small planes should be sent instead of one big one as the area has only a small airport.

For translation

It is very difficult for computers to translate from one language to another, as words can have different meanings in different sentences. For instance, the computer needs to be given a lot of information before it can recognize the difference between "I feel like a cup of tea" and "I feel like an idiot". Computers which can translate are being developed, though, using scanning devices as shown above, to "read" foreign scripts.

Programmable car

This toy car is controlled by a microprocessor chip and can be programmed to go round furniture, out of the room, and come back to you again. You program the car by tapping your instructions into the keyboard under the bonnet.

KEYBOARD

Making music

An artist, John Lifton, has used a computer to make music from plants. All plants contain tiny charges of electricity, and changes in these charges were recorded by the computer. The computer then sent messages to a sound synthesizer which made different sounds depending on the changes in the plants. People even brought their own plants to test on the computer to see what sounds they made.

Can computers think?

The computers and robots of science fiction can do everything humans can – and more. Present day machines are not quite so clever, though sometimes their responses do seem almost human, as in the conversation with a computer shown below. This sort of behaviour by a machine is called artificial intelligence.

Even these machines, though, are entirely controlled by the instructions in their programs. The main difficulty in creating machines which can think is in writing the programs for them.

A computer has to be programmed to recognize all the information it is given before it can act upon it. For instance, in order to read handwriting, it would have to be programmed to recognize all these shapes of "A", and many more.

For a computer to be able to think and act in an intelligent way, it would have to be programmed to recognize a vast amount of information, and to make decisions based on other information stored in its memory.

Conversation with a computer

These pictures illustrate a conversation with a computer. The person types the questions into the keyboard and the computer's answers appear on the screen.

The computer is programmed to recognize certain letters, words and signs, and to reply to them. The program also contains a list of ideas and grammar rules which the computer uses to work out its replies.

Once you know how this program works, the computer does not seem so clever. If we knew how our brains worked, though, would that make us seem less intelligent?

1 WHO ARE YOU? / I AM A PROGRAMMED COMPUTER.

The computer recognizes "O ARE Y" and "?" and these trigger off its reply.

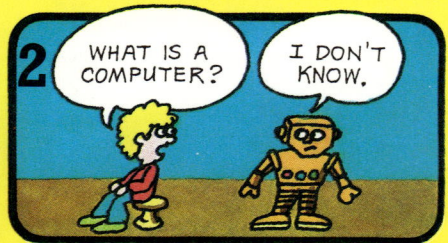

2 WHAT IS A COMPUTER? / I DON'T KNOW.

It spots "WHAT IS" and "?" and its reply is one of many it can give to this query. It also spots the word "COMPUTER", but it does not always react to every word in the sentence.

3 YOU MUST KNOW. / KNOWLEDGE IS A PROBLEM FOR ME.

This time the computer spots "YOU" and "KNOW", and gives a standard reply to the word "KNOW".

4 I THINK YOU ARE STUPID. / YOU THINK COMPUTERS ARE STUPID? / SOMETIMES. / I SEE.

In reply to the first statement the computer replaces the words "I" and "YOU" with "YOU" and "COMPUTERS". It often does this in a sentence of this kind. It does not recognize "SOMETIMES" so it gives a vague reply.

How computers play chess

The program for the computer contains a list of rules about the game. From these, the computer can work out all the moves which are possible at each stage in the game. It is mainly the speed with which it can examine each move, and how it will affect the game, which enables it to win.

Micromouse

Micromouse is a microprocessor on wheels which can find its way round a maze. Its actions are more like instinct than intelligence as it cannot plan its route. It "feels" round the walls with sensitive infra-red pulses and bendy wires, and is programmed to react to obstacles.

"Senster" the gentle robot

This is Senster, a robot made in 1971 by a British sculptor and engineer called Edward Ihnatowicz. It has been dismantled now.

A robot is a machine which can perform some of the actions of a human being or animal. Senster could twist its long metal neck to "look" at people and shied away from loud noises or sudden movements made by the people watching it. It was built as a piece of art.

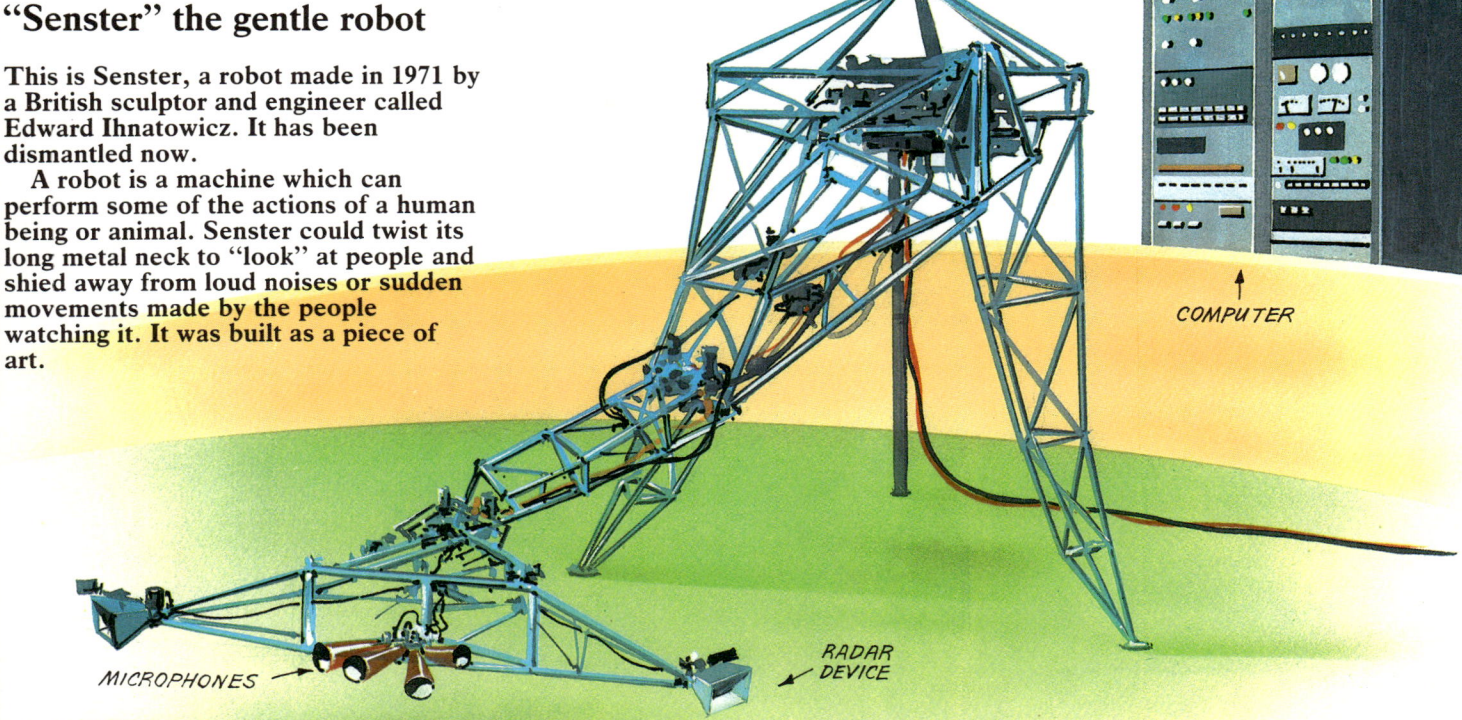

MICROPHONES

RADAR DEVICE

COMPUTER

Senster had four microphones with which it could hear noises, and radar devices to detect movements. It was controlled by a computer which was programmed to make the robot swing its head towards and away from noises and movements, depending on the signals from the microphones and radar.

The robot's radar and microphones were like our eyes and ears which give our brains information about what is happening around us. Human beings, though, can react to this information in a great number of ways, whereas a robot's responses are limited by the instructions in its program.

Robots can have many different electronic senses. As well as "sight" and "hearing", they can have a sense of "touch" via pressure pads which show how tightly they are gripping something, and can also sense heat and cold. A robot could never have all the skills and judgement of a human being, though.

Brain versus computer

Some people think it is only a question of time until a computer can be programmed to function like a human brain, while others believe it is a fantasy of science fiction stories. Here are some of the features of a human brain compared with a present day computer.

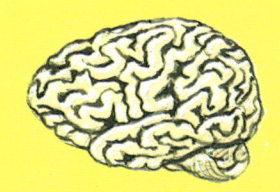

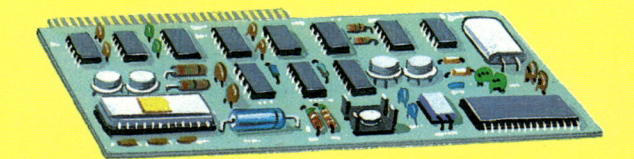

Brain

- Weighs about 1.5kg.
- Energy source is blood glucose.
- Needs a steady temperature.
- Number of basic computing elements estimated at about a hundred thousand million.
- The various parts of the brain must stay in one place.
- Rapid memory recall, apparently unlimited because of the way the brain stores ideas.
- Average intelligence rated at an IQ level of about 100.

Computer

- Weighs from a few grammes to tonnes.
- Energy source is electricity. (The amount of power a computer needs is getting less each year.)
- Less sensitive to heat and cold.
- At best, about a thousand million basic computing elements, but increasing all the time.
- The various parts can be in different places and linked by wires, satellites, laser beams, etc.
- Access speed to memory store limited by present technology.
- General intelligence about that of a very stupid worm.

19

Computers in offices and factories

Keeping business records, stock lists, sales accounts, and details of wages and pensions are all obvious ways in which computers can help people in business and industry.

They can also be used to help managers plan and make decisions, by presenting the information they need in ways that make it easier to understand. They can work out the best way to do complicated jobs, and even control the machinery that does the work. In all these cases, though, the computers still need skilled workers to program and control them.

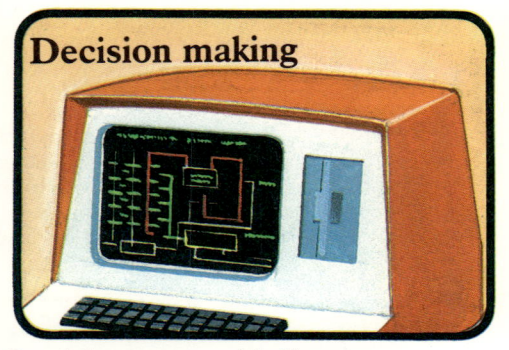

Decision making

Special programs with diagrams showing various courses of action and their probable results, can help planners making decisions. The programs can reveal unforeseen problems, and show how a decision may affect the whole company.

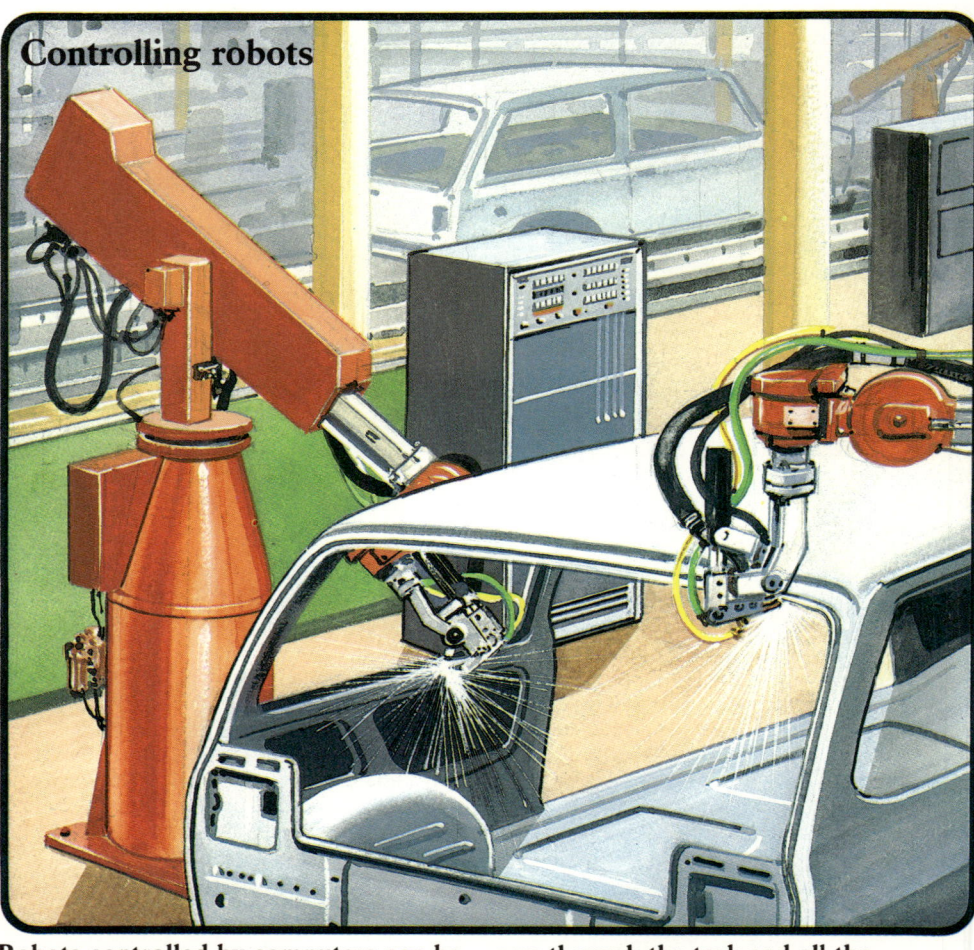

Controlling robots

Robots controlled by computers can be used instead of people to do boring or dangerous jobs. These robots are welding the joints on car frames. All their actions are controlled by a program of instructions in the computer.

Some robots are "taught" their jobs by people. A worker guides the robot's arm through the task and all the movements are stored as a program in the computer's memory. Then, when the program is rerun, the robot repeats the movements exactly. If the pattern of work changes, the robot has to be re-programmed, and it cannot cope with unexpected situations as a human worker could.

Solving problems

This is a computer-controlled machine for cutting the fabric for clothes. Using a computer, it is much easier for a skilled pattern cutter to work out how to position the pattern pieces so that as little fabric is wasted as possible. The lay-out of the pieces is stored in the computer program which is used to guide the cutting machine.

Giving information

In large production plants, computers can be used to give continual information about all the production processes. Some programs use simple symbols, like those on the screen above, which production controllers can understand at a glance. Each symbol represents a different process, such as the flow of raw materials and the state of machinery and furnaces.

Screen information

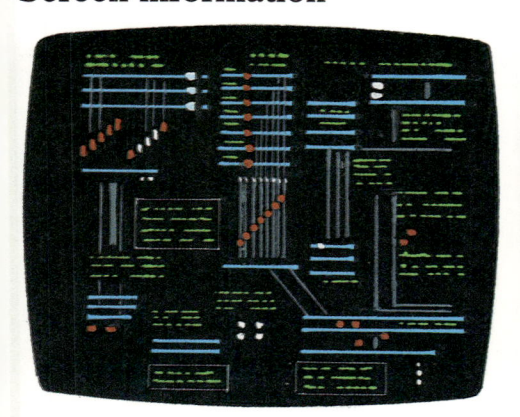

A screen full of facts and figures can be very difficult to understand. The computer can show the same information in lots of different ways, though, using words, colours and symbols which make it easier to understand.

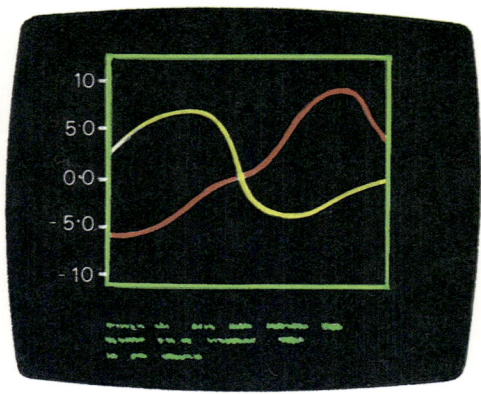

It can be programmed to show the same facts on a graph, using several different colours to make it clearer. If there is a lot of information, though, graphs can be difficult to read.

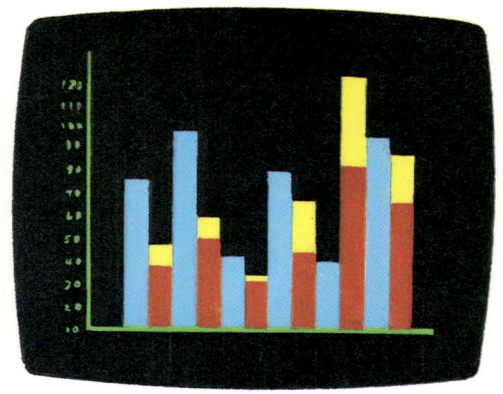

Bar charts and pie charts are easier to understand, but they cannot show very much detail. The person writing the program has to choose the best way for the computer to show the facts.

Another way to show the facts

These pictures show another, new way that computers can present information, in this case about road haulage companies. The computer creates pictures like these from facts about the companies, and the sizes of the various parts of the trucks show how well each company is doing.

For instance, the size of the load shows how much work they do, the wheels show how efficient they are, and the fumes show how much time and resources they waste. This way of displaying information makes it very easy to compare several sets of facts.

Make your own information diagrams

Size of goal = no. of games played this season
Size of ball = no. of games won
Size of scorer = no. of goals scored
Size of goalkeeper = no. of goals scored against team

THIS TEAM PLAYED AND WON LOTS OF GAMES, SCORED A LOT OF GOALS AND HAD QUITE A FEW SCORED AGAINST THEM.

THIS TEAM PLAYED FEWER GAMES, BUT WON MOST OF THEM, SCORED QUITE A FEW GOALS AND HAD FEW SCORED AGAINST THEM.

Here are some ideas for making information diagrams about football teams, to compare how successful they are. Each team is represented by a goal, and the size of the goal, goalkeeper, scorer and ball show how many games each team played, how many games they won, the number of goals scored, etc. You could make up your own rules, though, or think of other things to make diagrams of. For instance, you could compare two pop groups (number of records produced, number of hits, place in the charts), or motorcycles (top speed, fuel consumption, engine size, etc.)

Home computers

In the past, only highly trained computer experts were able to use computers, but as they become cheaper, smaller and easier to operate, many people such as doctors, teachers, librarians, architects and artists are finding that a computer can help them.

There are lots of uses, too, for a computer at home – to look after home finances, store useful information such as addresses and timetables and play games with. You can already buy a pocket-sized computer quite cheaply, and learn to use it in a few hours. Computer enthusiasts can even build their own computer from a kit.

1 This is an American "Altair", one of the first small "home" computers which was on sale in 1976. It could be built from a kit, but it was quite complicated to use and had lots of switches and flashing lights.

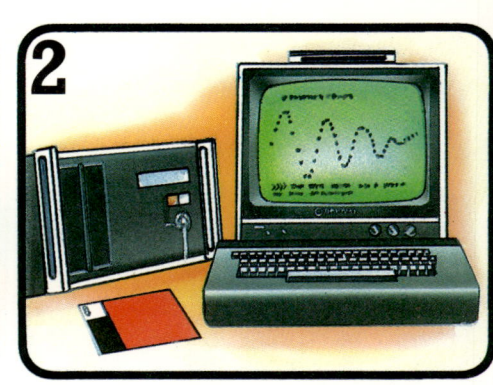

2 Within a few years, though, home computers have become much easier to use. They have very few buttons and switches and all the operating instructions and information are shown on the computer screen.

Computerized homes

Sometime in the future, houses will probably have built-in computers which control everything from paying the bills to opening the doors. An experimental house like this has already been built in Arizona, USA. This picture shows some of the things such a home computer could do.

The computer would control all the lights in the house. Sensors would record your movements as you went from room to room and the computer would switch the lights on and off. This would help save energy.

Hidden computers

Each of these things has a small computer inside, in the form of a microprocessor chip programmed to control how it works.

Microprocessors are used in many other kinds of equipment, such as washing machines, cookers, telephones and electronic games.

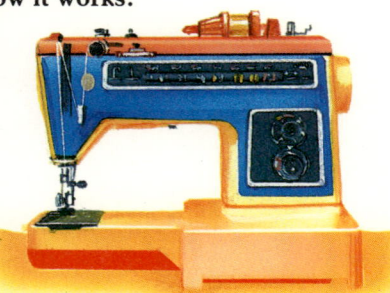

This sewing machine has a chip programmed to do lots of different embroidery stitches. To change the stitch you press a button and the chip sends instructions to the mechanical stitching parts of the machine.

The aperture and shutter speed of this camera are controlled by a microprocessor. Light sensors register how bright the light is, and the chip selects the correct aperture size and shutter speed.

This electronic keyboard can make the sounds of 29 musical instruments. It contains information about the sounds of the instruments in the memory of its chips. When you play on the keyboard, it makes the sound of the instrument of your choice.

3

Now that computers are easy to use, people can experiment with a computer and do their own programming. A few years ago, only computer experts had access to the computers, and they controlled how they worked.

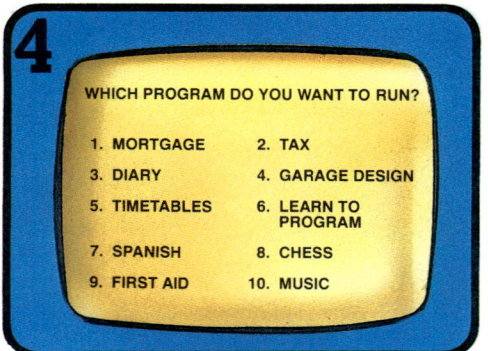

4

WHICH PROGRAM DO YOU WANT TO RUN?
1. MORTGAGE
2. TAX
3. DIARY
4. GARAGE DESIGN
5. TIMETABLES
6. LEARN TO PROGRAM
7. SPANISH
8. CHESS
9. FIRST AID
10. MUSIC

You can also buy pre-recorded computer programs on discs and cassettes. There are programs for computer games and for calculating bills, and also to help you learn a new language, or design a do-it-yourself project.

5

Microcomputers in schools and public libraries give lots of people the chance to use a computer. In the future, though, owning a computer will probably be as common as owning a wristwatch.

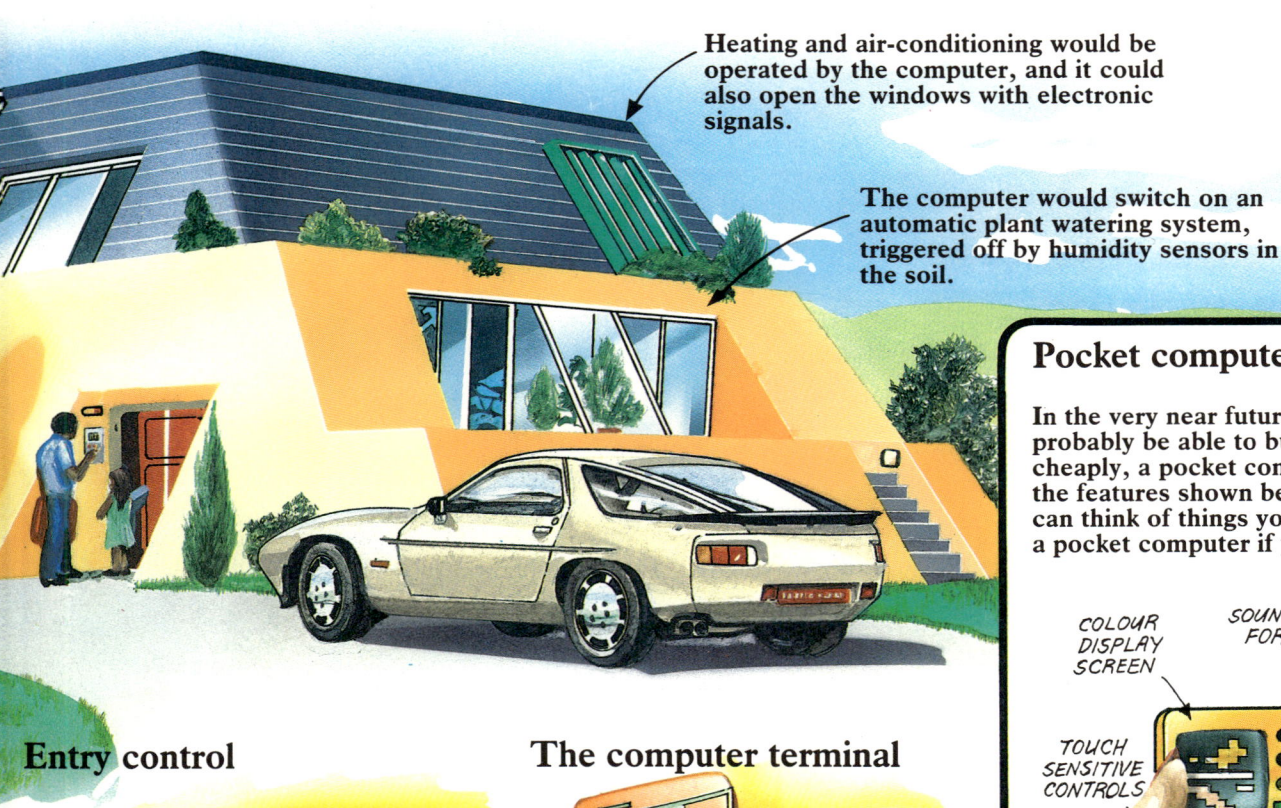

Heating and air-conditioning would be operated by the computer, and it could also open the windows with electronic signals.

The computer would switch on an automatic plant watering system, triggered off by humidity sensors in the soil.

Pocket computer

In the very near future you will probably be able to buy, quite cheaply, a pocket computer with all the features shown below. See if you can think of things you would do with a pocket computer if you had one.

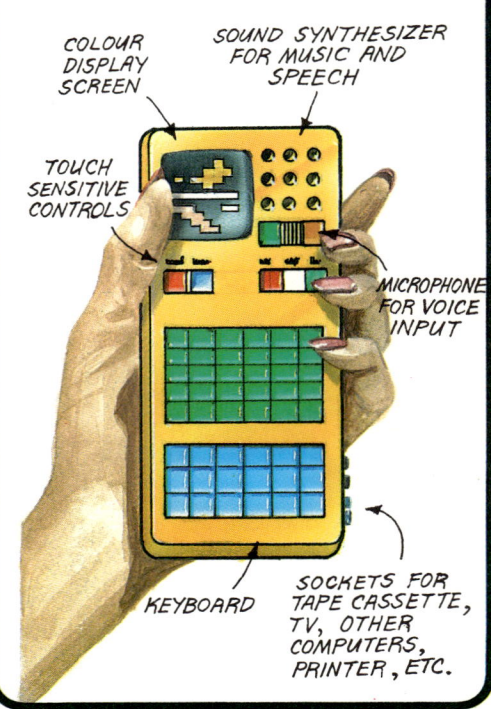

COLOUR DISPLAY SCREEN
SOUND SYNTHESIZER FOR MUSIC AND SPEECH
TOUCH SENSITIVE CONTROLS
MICROPHONE FOR VOICE INPUT
KEYBOARD
SOCKETS FOR TAPE CASSETTE, TV, OTHER COMPUTERS, PRINTER, ETC.

Entry control

To enter the house you would punch your personal code into a keyboard on the door. The computer could be programmed to let in certain people only, and could take messages and reply in a human-like voice, using a speech synthesizer.

The computer terminal

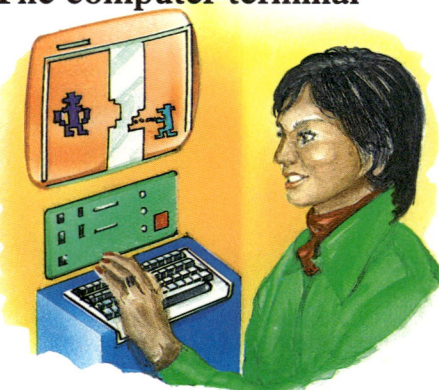

The computer itself could also be used for many other purposes: for storing information, paying bills, playing games, in fact, for anything you programmed it to do.

23

Instant information

A large computer can search through its databanks and find any bit of information within a few seconds. It is even possible to call up a computer hundreds of miles away and receive information from it on your own computer – or even on your television set. These two pages show some of the ways that computers can be linked together, and how the information in a databank can be used, and misused.

Databank information, and programs and their results can be sent from one computer to another via a satellite. The computer signals are converted to high frequency radio waves, which are beamed up to the satellite, then reflected back to the receiving computer and changed back into computer signals again. Computer data can also be sent from computer to computer along the telephone cables.

Sharing a computer

RESEARCH LABORATORIES

SCHOOLS

HOME COMPUTERS

OFFICES

A large computer can be used by lots of people at the same time. In fact, the computer deals with the data from each person separately, but it switches from one to the other so fast that no-one notices the delay.

In this way, lots of organizations can share the expensive resources of a large mainframe computer and its databank. They can be linked to the computer by direct cables, by telephone cables, or even by satellite. Each user has a "user number" and a password which entitles them to use the computer.

Databanks

The ability of a computer to sort through the vast amounts of information in its databank, and recognize and show us relevant items, is of great benefit to research in science and medicine. For instance, if doctors had a databank of all the known diseases and their symptoms, they would be able to diagnose illnesses much more easily.

People are concerned, though, about the amount of personal information about people which is held in databanks. Such information, in the wrong hands, could easily be used to discriminate against people. Some countries have laws controlling the use of databanks and allowing people to check the information about themselves.

24

TV sets as computer terminals

In the near future, television sets will probably be used for much more than watching TV programmes. Already you can use a TV set as a display screen for a home computer, and for computer games. It is also possible to receive information from a central computer on your TV set. The two different ways in which the television can receive information from the computer are shown here.

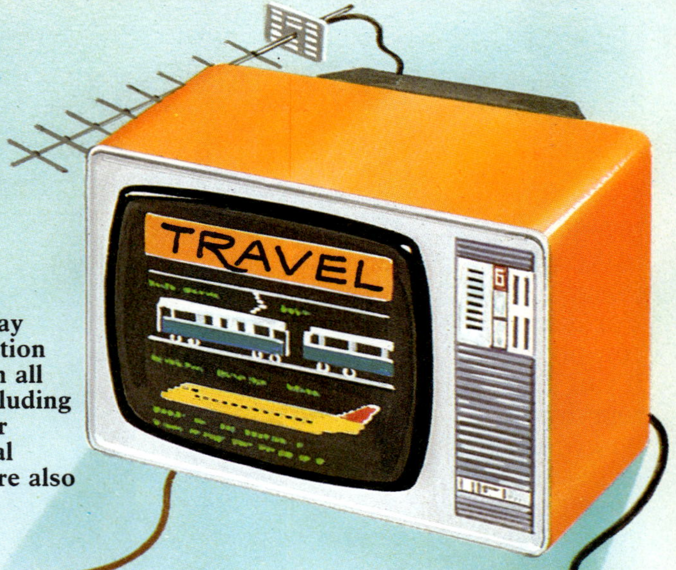

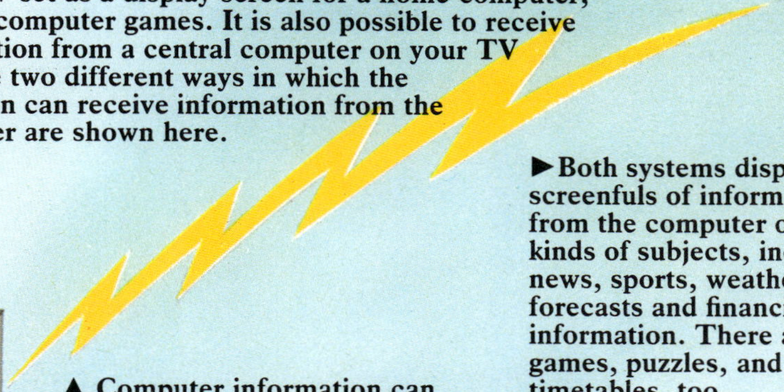

▶ Both systems display screenfuls of information from the computer on all kinds of subjects, including news, sports, weather forecasts and financial information. There are also games, puzzles, and timetables, too.

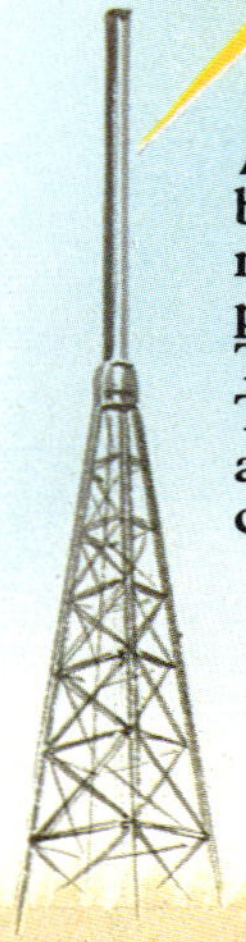

▲ Computer information can be transmitted along with the normal television signals, and picked up by the TV aerial. This system is called teletext. The television set has to be adapted to receive the computer signals.

◀ For both systems you choose the information you want with a hand-held keypad.

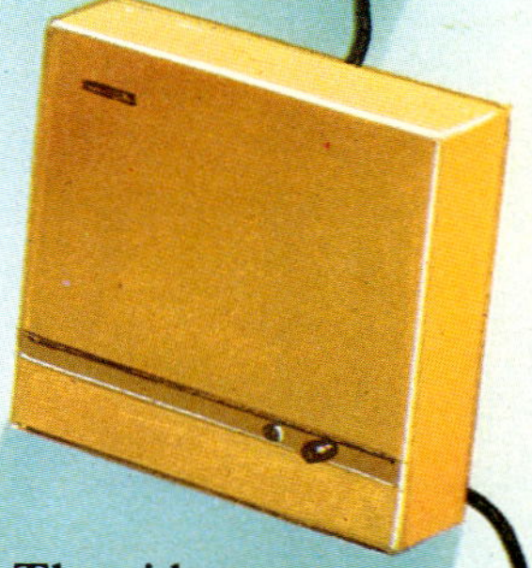

◀ The information is stored in a central computer. Several different companies transmit information by teletext and videotex, and each has its own computer.

▲ The videotex signals have to be decoded in a special unit called a modem before the TV can understand them.

◀ Another system, called videotex, sends information from a central computer as signals along the telephone wires. The telephone itself, though, plays no part in the system.

Computer crimes

Nowadays, with so much information passing between computers, including orders for the delivery of goods, and instructions for payments, a new kind of crime has developed. People have discovered that they can intercept computer messages and alter the information. To try and prevent this, computers are programmed to accept new instructions only if they receive certain codes and passwords. Computer criminals still find ways to send false messages though.

Recently, in the USA some boys managed to tap into the computer of a company that sold computer components. They told the computer to deliver parts to them until they had enough to build their own computer – but then they were caught.

Computer models

Sometimes it is easier to solve a problem, or test an idea, if you make a model of it. You can use this method with a computer, too. For instance, if you give the computer details about an airport, it can put all the information together and provide a simple description or "model" of the airport and how it works. The model is presented as words, charts, graphs and pictures, it is not a solid, 3D model.

You can then give the computer new information, and tell it to show how this changes the model. Using computer models like this is called computer simulation.

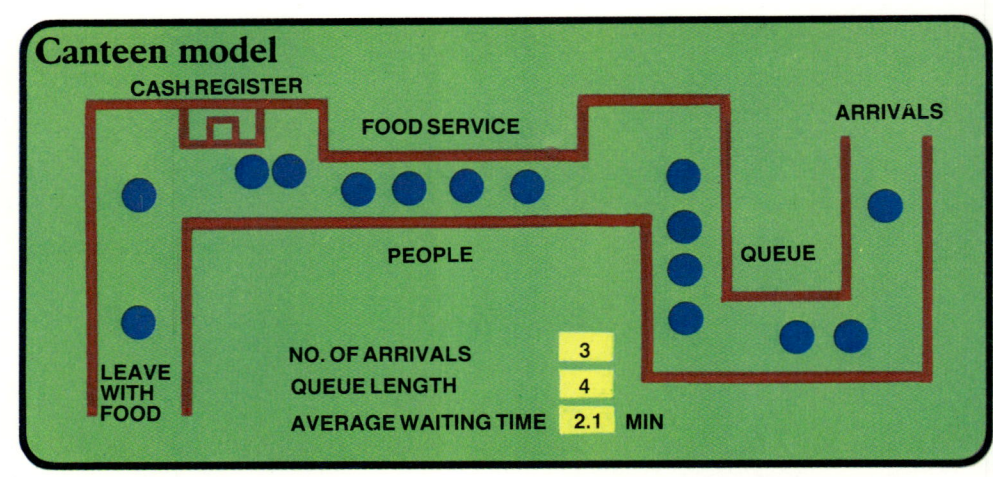

Canteen model

This is a computer model of how a queue builds up in a canteen. To make the model, the computer was given information about when people arrive, how fast they are served, the length of the queue and delays at the cash register.

Then, given new information, the computer can show what happens to the queue if the rate of arrival of people changes, or the food is served faster. It can work out how fast everything should be so there are no delays, and food does not get cold.

Ship's bridge simulator

Computer simulations are used to help people learn to navigate large "supertankers". A room is fitted out like the bridge of a tanker, and the view through the windows is created by a computer. The tanker's instruments and controls are connected to the computer, which, as the "captain" steers the ship, alters the view through the windows. In this way, a trainee can learn to steer the tanker without causing any harm if the ship runs aground or is in a collision. Aircraft pilots learn to fly in cockpits with computer simulated views too.

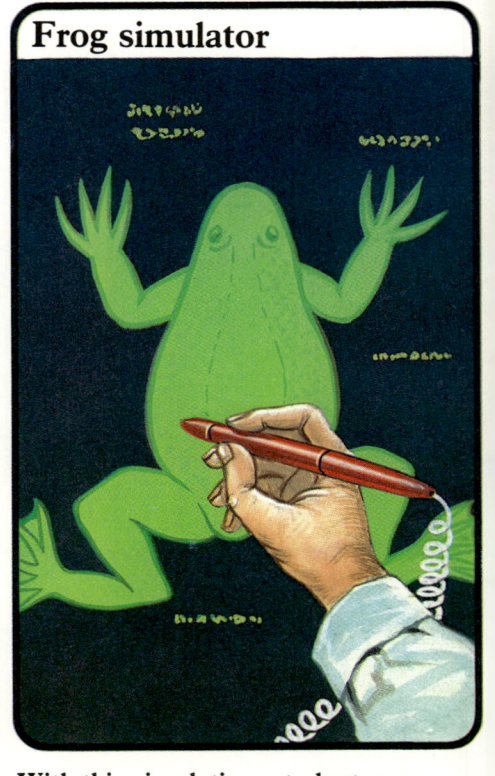

Frog simulator

With this simulation, students can learn how to dissect a frog without cutting up a real one. Using a light sensitive pen, the student "dissects" a picture of a frog created by a computer on a display screen. The computer program contains all the information about the frog's organs and tissues, and each time the student makes a "cut" the computer changes the picture to show the frog's insides.

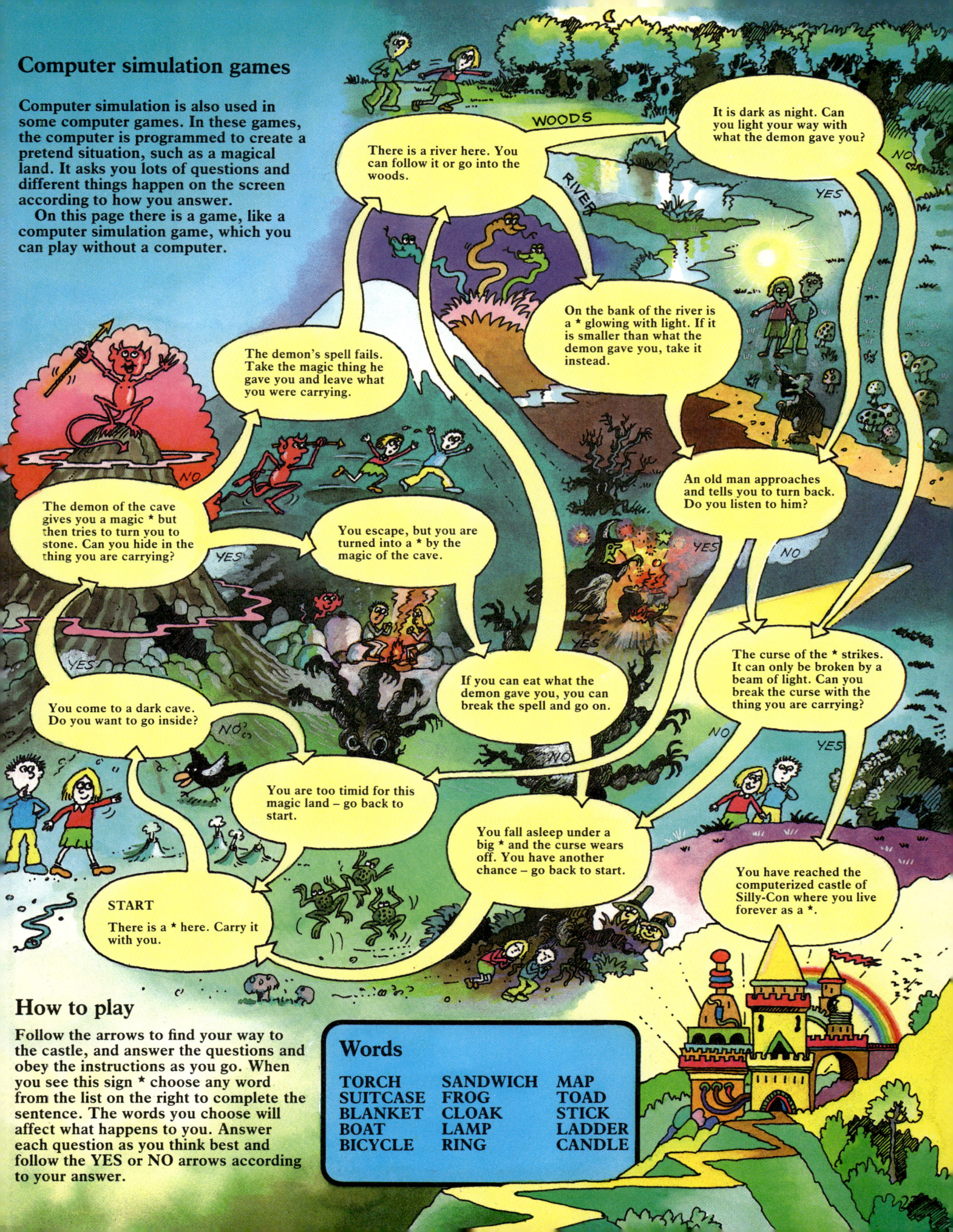

Computer simulation games

Computer simulation is also used in some computer games. In these games, the computer is programmed to create a pretend situation, such as a magical land. It asks you lots of questions and different things happen on the screen according to how you answer.

On this page there is a game, like a computer simulation game, which you can play without a computer.

There is a river here. You can follow it or go into the woods.

It is dark as night. Can you light your way with what the demon gave you?

On the bank of the river is a * glowing with light. If it is smaller than what the demon gave you, take it instead.

The demon's spell fails. Take the magic thing he gave you and leave what you were carrying.

An old man approaches and tells you to turn back. Do you listen to him?

The demon of the cave gives you a magic * but then tries to turn you to stone. Can you hide in the thing you are carrying?

You escape, but you are turned into a * by the magic of the cave.

The curse of the * strikes. It can only be broken by a beam of light. Can you break the curse with the thing you are carrying?

You come to a dark cave. Do you want to go inside?

If you can eat what the demon gave you, you can break the spell and go on.

You are too timid for this magic land – go back to start.

You fall asleep under a big * and the curse wears off. You have another chance – go back to start.

You have reached the computerized castle of Silly-Con where you live forever as a *.

START
There is a * here. Carry it with you.

How to play

Follow the arrows to find your way to the castle, and answer the questions and obey the instructions as you go. When you see this sign * choose any word from the list on the right to complete the sentence. The words you choose will affect what happens to you. Answer each question as you think best and follow the YES or NO arrows according to your answer.

Words

TORCH	SANDWICH	MAP
SUITCASE	FROG	TOAD
BLANKET	CLOAK	STICK
BOAT	LAMP	LADDER
BICYCLE	RING	CANDLE

Computers and art

Artists can use computers to help them create pictures, patterns and sculptures, and even music and poetry too. Lots of the special effects in films and on television are created with computers, and they are also used to make cartoon films.

In the past, most computer art was not very successful as the computers had to be programmed by computer scientists who were not necessarily very good artists. Artists can now do their own programming, and experiment with computers in their work.

Drawing robot

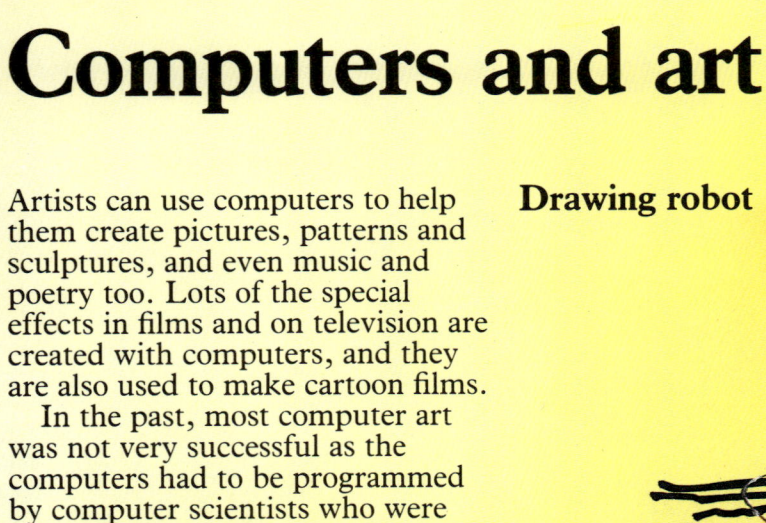

WIRE TO COMPUTER

ROBOT CONTROLLED BY COMPUTER

An artist called Harold Cohen, has used a robot controlled by a computer to make abstract drawings like this. The robot rolls across the paper, raising and lowering a pen underneath it according to the instructions in the program. Cohen tells the computer which patterns he likes best, and the computer stores them in its memory. In this way the robot's drawings improve as the computer builds up a store of successful patterns and forgets the ones which do not work so well. When the program is rerun, the robot can reproduce the patterns stored in the computer's memory exactly.

Cartoon films

A cartoon film is made up of thousands of pictures showing the characters in lots of slightly different positions. When the film is shown at the normal speed, the figures look as though they are moving. A computer can be programmed to work out all the pictures much more quickly than an artist can. Given the positions of the legs at the beginning and end of a walking sequence, the computer can work out all the leg positions in between, and display them on a screen or draw them out on a plotter. This is called "inbetweening".

1. How a computer makes screen pictures

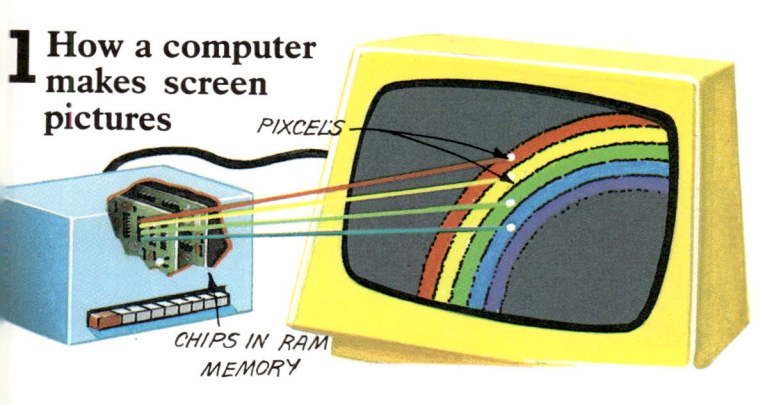

The screen is divided into tiny squares called "pixcels" and the pattern of pixcels which are turned on make up the picture. Each pixcel is controlled by a small portion of the computer's RAM memory. You give the computer the information for the picture by typing on the keyboard, drawing on a sensitive board, or by scanning a picture with a special camera.

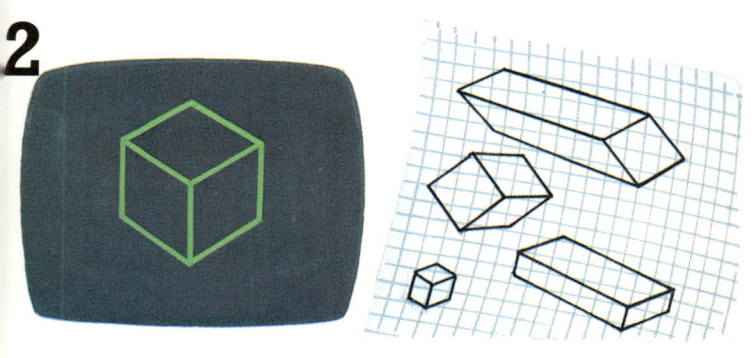

2.

The information for the picture is stored in binary code in the computer's memory and it is very easy to alter the information, and so change the picture, by typing in new instructions. For instance, you can change the colour or shape of a picture and, as with the cube shown here, enlarge, reduce, stretch or rotate it. By storing the program for the picture on a disc or tape, you can get exactly the same picture every time you run the program.

3.

The quality of the pictures depends on how many pixcels there are in the screen, and how much memory the computer has, for each pixcel needs its own little bit of memory. Realistic pictures, like the one on the left are made up of many thousands of tiny pixcels. More stylized pictures like the one on the right have fewer, larger pixcels and so need less computer memory.

Computer paintings

This artist is making a picture on a computer screen by drawing on a sensitive board with a special pen. The artist can use any of the colours shown along the bottom of the screen, and tells the computer which colour to use by touching the place for that colour in the instruction grid along the top of the board.

After the picture is finished, the artist can change it if he likes by telling the computer to make it larger or smaller, alter the viewpoint or even change the colours. Then the picture can be stored on a magnetic disc. Pictures are often produced for television like this, and they can be transmitted almost immediately.

Computer flick book

You could make a flick book to see how the walking sequence on the left works. To make it, fold a large piece of paper over and over until it measures about 7cm x 5cm. Staple or stitch it along one side, then trim the other edges to make the pages. Trace all the leg pictures shown on the left and copy one pair of legs on to every page, putting them in the same position on each page.

Computer firsts

The development of computers can be divided into three main stages, or generations. The first generation was the large mainframes built with valves. The smaller more reliable computers built with transistors are called the second generation, and computers made with silicon chips are the third generation. Here is a list of the main dates in the history of computers.

1945 ENIAC, the first all-electronic machine, was built. It was more like a calculator than a present-day computer, though, as it could not store data or programs. ENIAC stands for Electronic Numerical Integrator and Calculator.

1947 A new kind of electronic component, called the transistor, was invented. Transistors were first used in computers in about 1953.

1948 The Manchester University Mark I, the first real computer (that is, one which could store a program of instructions), ran for 52 minutes on June 21.

1950 The Ferranti Mark I, based on the Manchester Mark I, was sold commercially in Europe.

1958 The first working integrated circuit was developed.

1960 The first "chips" – integrated circuits on chips of silicon – were produced.

1964 The first computers built with integrated circuits were produced for the general market.

1975/6 The first small "home" computer, the "Altair", was on sale.

1980 The first pocket computer, the Japanese Sharp PC1211 was sold.

The future

This graph shows how dramatically computers have developed over the last 40 years, becoming smaller, cheaper and more and more powerful. The first computers could do relatively few calculations a second whereas a 1980 mainframe can carry out millions of instructions each second. These trends will probably continue and perhaps the only restrictions on the future of computers will be the cost of new inventions and whether we really want such machines. If cars had developed at the same rate as computers we would now be able to travel at thousands of kilometres an hour in tiny vehicles which would use hardly any fuel and be relatively cheap to buy.

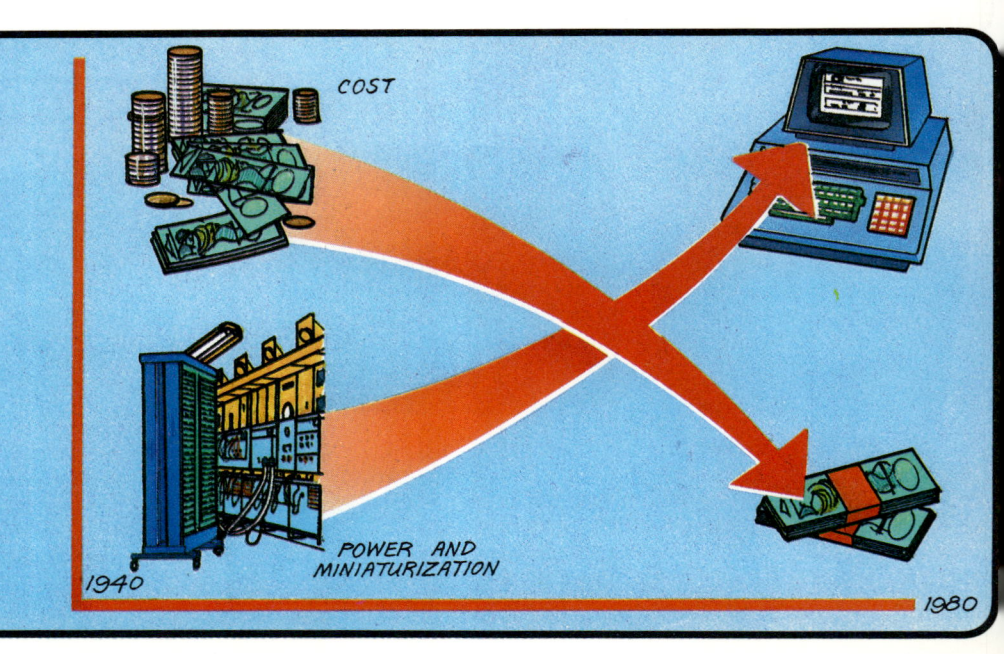

Computer words

Here is a list of computer words and their meanings. If you want to know more about one of the words, look it up in the index, then turn to the pages for that word.

BACKING STORE Information stored by the computer on magnetic discs, tapes, etc.

BINARY A number system based on the two digits "1" and "0".

BIT A binary digit, that is, "1" and "0".

BUG An error in a computer program. Bugs are usually a nuisance, but sometimes they are lucky accidents which open up new lines of approach to a problem.

BYTE A group of eight binary digits used to represent one unit of data in the computer's memory. The size of the computer's memory is measured in bytes.

COMPUTER A machine that can accept data, process it according to a stored program of instructions and then output the results.

CPU (CENTRAL PROCESSING UNIT) The control centre of the computer which organizes all the other parts inside the computer.

DATA The information given to the computer for processing.

DATABANK A very large quantity of information stored on magnetic discs and tapes, which the computer can sort through very quickly.

FLOWCHART A chart showing the sequence of steps needed in the program for solving a problem with a computer.

HARDWARE All the computer equipment, including the computer itself, input and output equipment, and magnetic disc and tape equipment.

INPUT The data going into the computer and the process of putting it in.

MEMORY The chips in the computer where information and instructions are stored in binary code.

MICROPROCESSOR A silicon chip which can do the same jobs as the main parts inside a computer.

OUTPUT The results of the computer's processing.

PRINTER Device which can print out the information from a computer.

PROGRAM A list of instructions, fed into or stored in the computer, which enable the computer to carry out a specific task.

PROGRAMMING LANGUAGE A special language, such as BASIC or PILOT, in which a program is written so the computer can understand it.

RAM (RANDOM ACCESS MEMORY) The part of the computer's memory where data, instructions and results are stored temporarily.

ROBOT Machine which can perform some of the movements of a person or animal.

ROM (READ ONLY MEMORY) The part of the computer's memory containing a permanent store of instructions for the computer.

SOFTWARE A collection of computer programs.

VDU (VISUAL DISPLAY UNIT) A TV-like screen on which information from the computer can be displayed.

Books to read

The four books on the left are simple, easy to read descriptions of the development of computers and how they work. The books on the right are more detailed accounts of computers, what they can do and how they affect the way we live.

The Challenge of the Chip by W. H. Mayall, HMSO, 1980 (obtainable from HMSO, 49 High Holborn, London WC1, England).
Computers by Heinz Kurth, World's Work, 1980
The Computer Age by Martin Campell-Kelly, Wayland, 1978
The Computer by David Carey, Ladybird Books, 1979

The Micro Revolution by Peter Laurie, Futura, 1980
The New Technology published by C.I.S., 1980 (obtainable from C.I.S., 9 Poland Street, London W1, England).
The Mighty Micro by Christopher Evans, Coronet, 1980

Buying a computer

One of the best ways to find out about computers is to own one yourself, and you can now buy a microcomputer for less than the price of a colour television set. With one of these small computers you can play endless games, write your own programs, do complicated calculations and even draw pictures and make electronic music. If you are thinking of buying your own computer, here are some guidelines to help you.

The hardware ▶
The simplest home computer consists of a keyboard which you can plug into your TV set. The keyboard contains all the chips which do the computer's work. You put the information into the computer by typing on the keyboard, and your programs and results are displayed on the TV screen. Some manufacturers also sell their computers as kits which you can assemble yourself. The kits are usually about 20% cheaper than the ready made computers.

Memory size ▶
Check how much RAM memory space the computer has. This is where the information you put into the computer is stored. The memory of a microcomputer is measured in kilobytes, written K. (One kilobyte = 1024 bytes.) Home computers usually have between 1K and 64K RAM. A memory of 1K is enough for learning to program and playing a few games, but as soon as you start writing your own programs you will need more memory. For most computers you can buy extra RAM memory to plug into the computer.

Backing store ▶
With most home computers you can store programs and data by recording them on a cassette tape using an ordinary cassette recorder.

Writing programs ▶
Most home computers use the programming language BASIC, but different makes of computer use slightly different versions of BASIC. The manual which comes with the computer shows you how to program it in BASIC.

◀ Software
You can also buy cassettes with games and other programs recorded on them for your computer. The programs for one computer do not usually work on another make of computer, so check that there is a good range of software for the computer you are interested in.

To find out more about the different kinds of home computer you can buy computer hobby magazines from newsagents. These frequently have test reports of popular machines. If you go along to a large computer store, the sales people will be willing to demonstrate the different machines for you, and will probably let you try them out too. Ask how much extra RAM costs for the different machines, and what extra equipment is available.

If there is a computer club near where you live you can probably go along as a guest and ask the members about their machines. To find the address of a club, try looking in your local paper, or ask in the library. Once you have bought your computer you will probably want to join the club and exchange programs with the other members. The computer hobby magazines also print lots of programs for home computers, especially games programs.

32 The word written in binary code on page 11 is ROBOT.

TV & VIDEO

Christopher Griffin-Beale and Robyn Gee

Contents

- 34 The TV revolution
- 36 Broadcasting
- 38 TV and video cameras
- 40 Colour cameras
- 41 Television sound
- 42 In a TV studio
- 44 In the control room
- 46 Transmitting TV pictures
- 48 TV sets
- 50 Outside the studios
- 52 Recording and editing
- 54 Special electronic effects
- 56 Digital effects
- 58 TV and computers
- 60 Home video equipment
- 62 Using a video camera
- 63 TV in the future
- 64 TV and video words

TV & Video was written by Christopher Griffin-Beale and Robyn Gee and illustrated by Ian Stephen, Graham Round, Graham Smith, Philip Schramm, Joe McEwan, Clifford Meadway and Martin Salisbury.
Designers: Graham Round, Kim Blundell and Roger Priddy

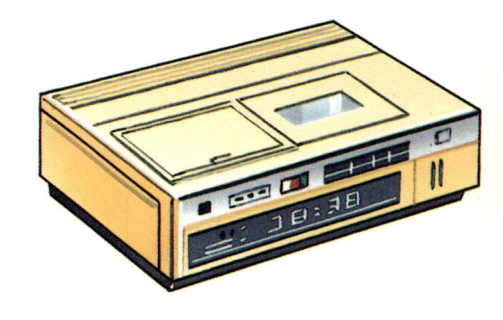

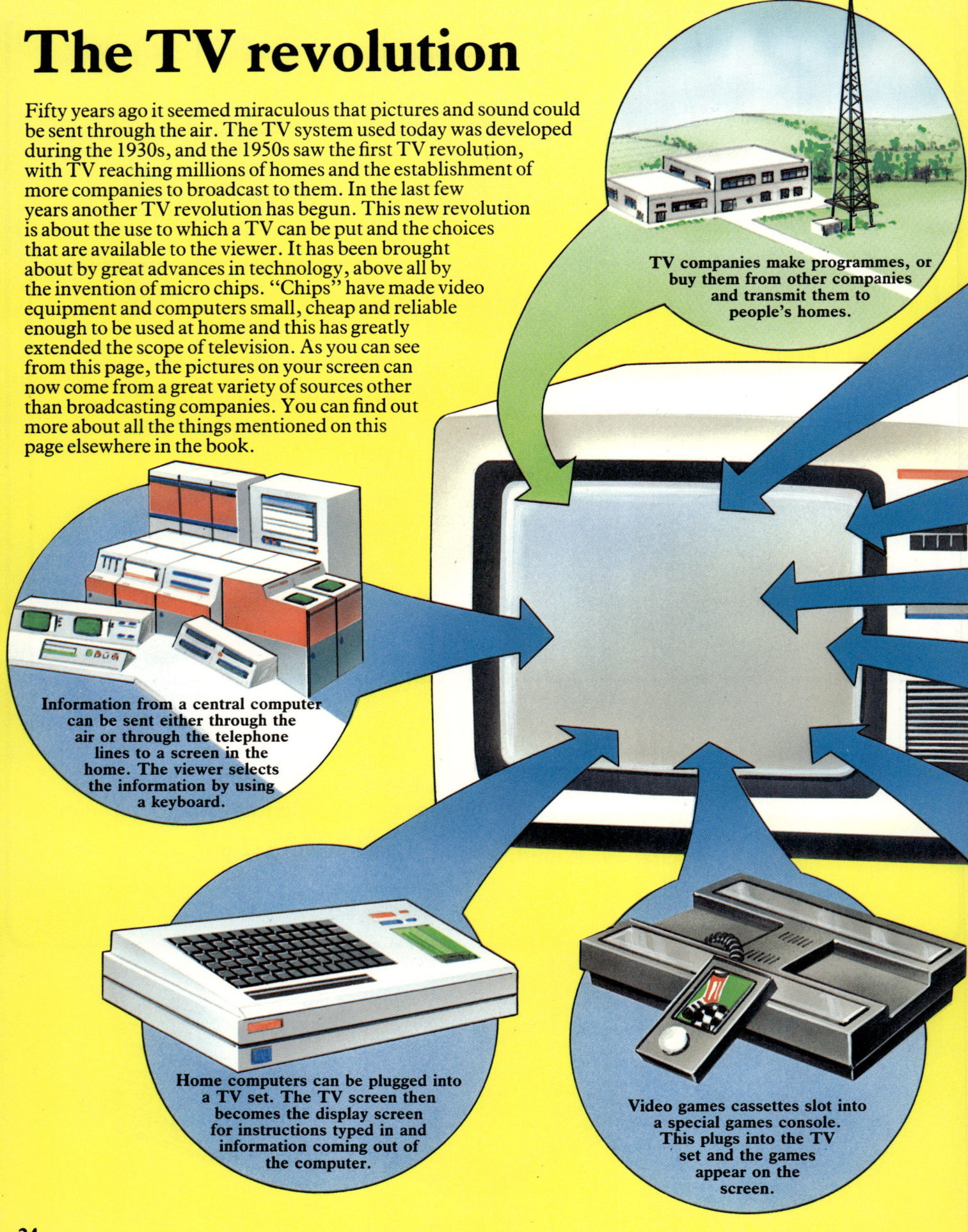

The TV revolution

Fifty years ago it seemed miraculous that pictures and sound could be sent through the air. The TV system used today was developed during the 1930s, and the 1950s saw the first TV revolution, with TV reaching millions of homes and the establishment of more companies to broadcast to them. In the last few years another TV revolution has begun. This new revolution is about the use to which a TV can be put and the choices that are available to the viewer. It has been brought about by great advances in technology, above all by the invention of micro chips. "Chips" have made video equipment and computers small, cheap and reliable enough to be used at home and this has greatly extended the scope of television. As you can see from this page, the pictures on your screen can now come from a great variety of sources other than broadcasting companies. You can find out more about all the things mentioned on this page elsewhere in the book.

TV companies make programmes, or buy them from other companies and transmit them to people's homes.

Information from a central computer can be sent either through the air or through the telephone lines to a screen in the home. The viewer selects the information by using a keyboard.

Home computers can be plugged into a TV set. The TV screen then becomes the display screen for instructions typed in and information coming out of the computer.

Video games cassettes slot into a special games console. This plugs into the TV set and the games appear on the screen.

34

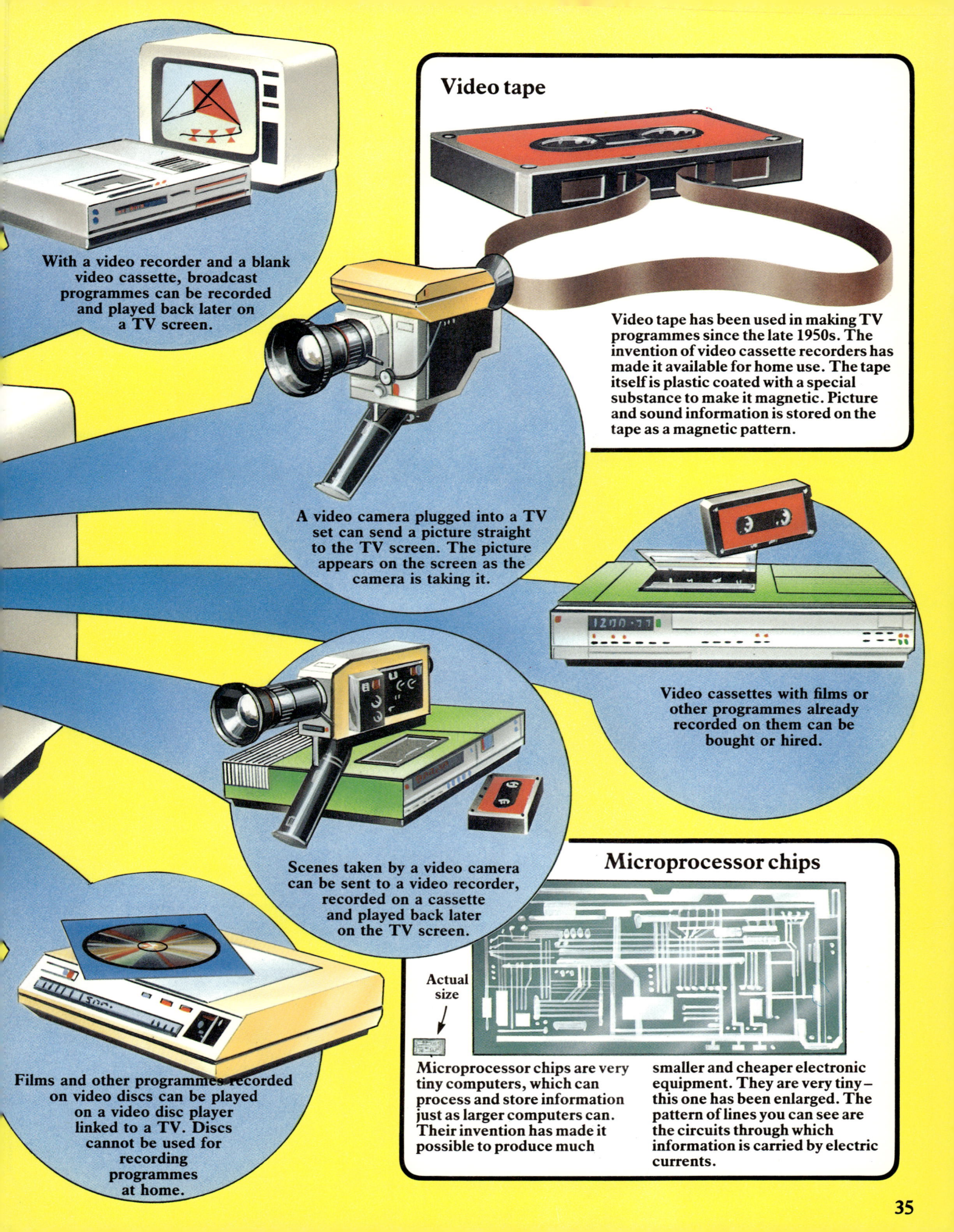

With a video recorder and a blank video cassette, broadcast programmes can be recorded and played back later on a TV screen.

Video tape

Video tape has been used in making TV programmes since the late 1950s. The invention of video cassette recorders has made it available for home use. The tape itself is plastic coated with a special substance to make it magnetic. Picture and sound information is stored on the tape as a magnetic pattern.

A video camera plugged into a TV set can send a picture straight to the TV screen. The picture appears on the screen as the camera is taking it.

Video cassettes with films or other programmes already recorded on them can be bought or hired.

Scenes taken by a video camera can be sent to a video recorder, recorded on a cassette and played back later on the TV screen.

Microprocessor chips

Actual size

Microprocessor chips are very tiny computers, which can process and store information just as larger computers can. Their invention has made it possible to produce much smaller and cheaper electronic equipment. They are very tiny – this one has been enlarged. The pattern of lines you can see are the circuits through which information is carried by electric currents.

Films and other programmes recorded on video discs can be played on a video disc player linked to a TV. Discs cannot be used for recording programmes at home.

35

Broadcasting

TV broadcasting is the process of sending out pictures and sound through the air. It developed out of two other important inventions: cinema photography (a way of capturing and reproducing moving images) and radio broadcasting (a way of sending sound through the air on radio waves). Once people had discovered how to adapt these two systems to make radio waves carry moving pictures as well as sound, "tele-vision" (literally "seeing at a distance") had been invented.

All the broadcast pictures you see on your TV screen are either "live", which means that the action is happening at the same time as you are seeing it, or "prerecorded", that is recorded before being broadcast. Prerecorded pictures are recorded either on film by a cine camera, or on videotape by a TV camera. Some programmes are made up of a mixture of live and prerecorded pictures.

Live programmes

In the early days of television all programmes were broadcast live. Now most are recorded before being broadcast. Those that still go out live are the ones where it is important to use the very latest information. Even live programmes are usually recorded during broadcasting, so that there is a copy to refer to or to show again.

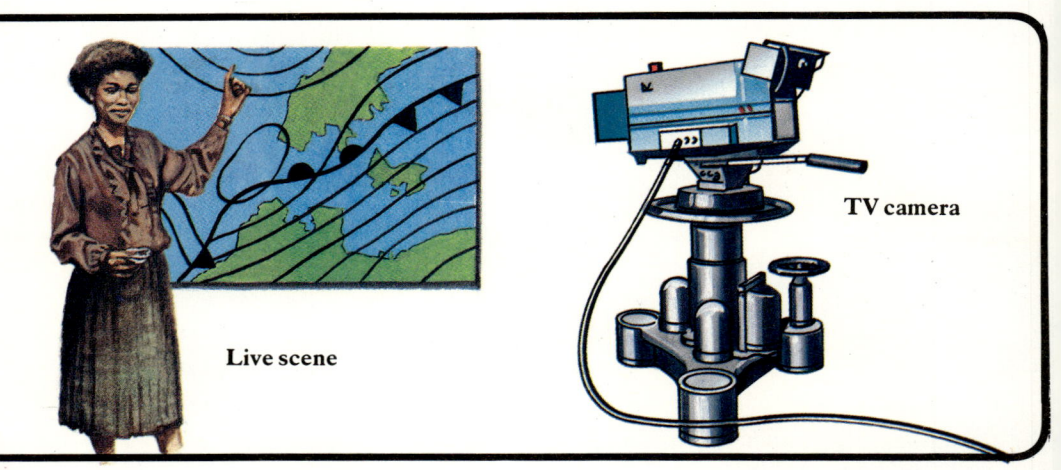

Live scene

TV camera

Prerecording on video tape

Video tape recording has been used in television for more than twenty years. Microphones and TV cameras send information to video tape recorders where it is stored on magnetic tape very similar to that used in sound recorders. Prerecorded tapes are stored in a video tape room ready to be played back when it is time to broadcast them. Mistakes can be cut out and new bits put in before it is broadcast.

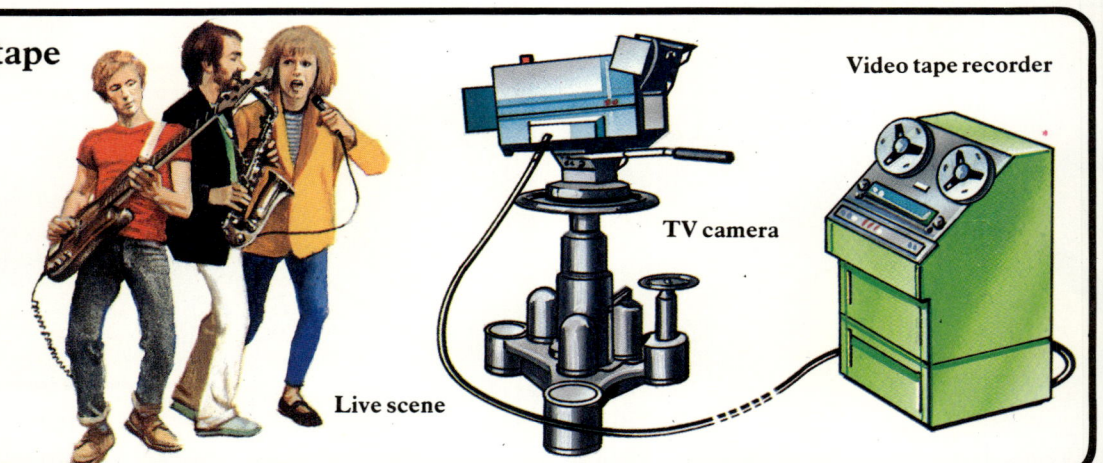

Live scene

TV camera

Video tape recorder

Prerecording on film

Some programmes are recorded on film. In this case the picture is recorded on plastic coated with a light sensitive chemical which has to be developed before it can be used. To broadcast material recorded on film, it has to be run through a telecine machine. This converts the film images into electronic signals which can then be broadcast in the same way as the electronic signals straight from TV cameras.

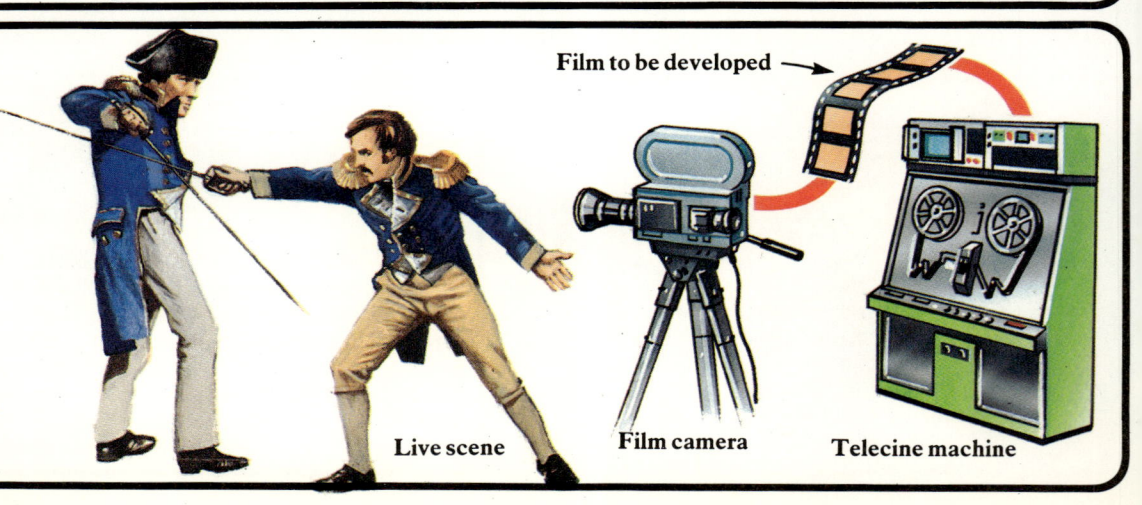

Live scene

Film camera

Film to be developed

Telecine machine

Stages in a TV broadcast

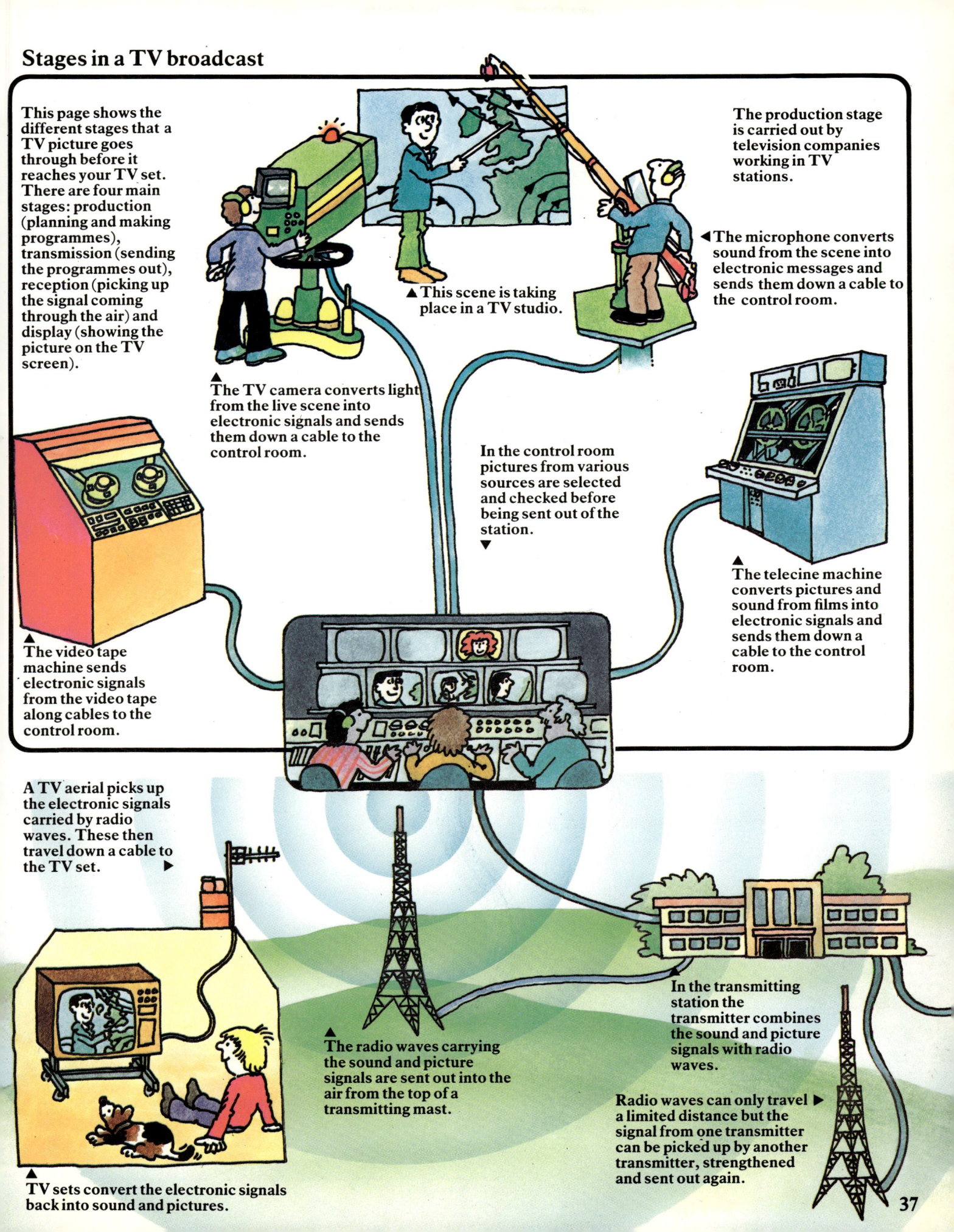

This page shows the different stages that a TV picture goes through before it reaches your TV set. There are four main stages: production (planning and making programmes), transmission (sending the programmes out), reception (picking up the signal coming through the air) and display (showing the picture on the TV screen).

The production stage is carried out by television companies working in TV stations.

◄ This scene is taking place in a TV studio.

◄ The microphone converts sound from the scene into electronic messages and sends them down a cable to the control room.

▲ The TV camera converts light from the live scene into electronic signals and sends them down a cable to the control room.

In the control room pictures from various sources are selected and checked before being sent out of the station. ▼

▲ The telecine machine converts pictures and sound from films into electronic signals and sends them down a cable to the control room.

▲ The video tape machine sends electronic signals from the video tape along cables to the control room.

A TV aerial picks up the electronic signals carried by radio waves. These then travel down a cable to the TV set. ▶

▲ The radio waves carrying the sound and picture signals are sent out into the air from the top of a transmitting mast.

In the transmitting station the transmitter combines the sound and picture signals with radio waves.

Radio waves can only travel ▶ a limited distance but the signal from one transmitter can be picked up by another transmitter, strengthened and sent out again.

▲ TV sets convert the electronic signals back into sound and pictures.

37

TV and video cameras

TV cameras are electronic. They work by turning light into electric currents, which are sent out of the cameras along electric cables. There is no film or tape inside them.

The electric currents can either be sent to a transmitter to be broadcast through the air on radio waves, or they can be sent to a video tape recorder to be stored and transmitted later. They are often transmitted and recorded at the same time.

A video camera works like a TV camera, but is smaller and lighter.

TV cameras come in various different types and sizes. They look quite like film cameras, but they are usually much bigger. The ones used in TV studios are often mounted on wheeled stands called dollies, so that they can be moved about easily.

Most TV cameras nowadays have a zoom lens. This allows the cameraman to alter his view from a long-shot to a close-up very fast. For the viewer this gives the impression that you are zooming towards the subject.

The picture taken by a TV camera can appear on a TV screen at almost the exact instant that it is being taken. The viewfiender of the camera is, in fact, a miniature TV screen. The picture that the cameraman sees there has already been taken and processed by the camera.

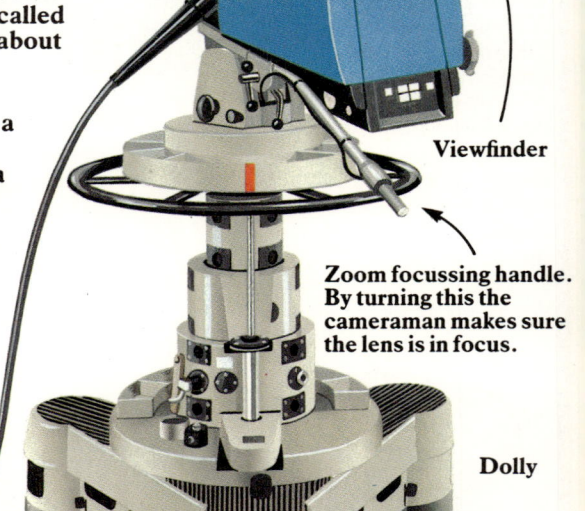

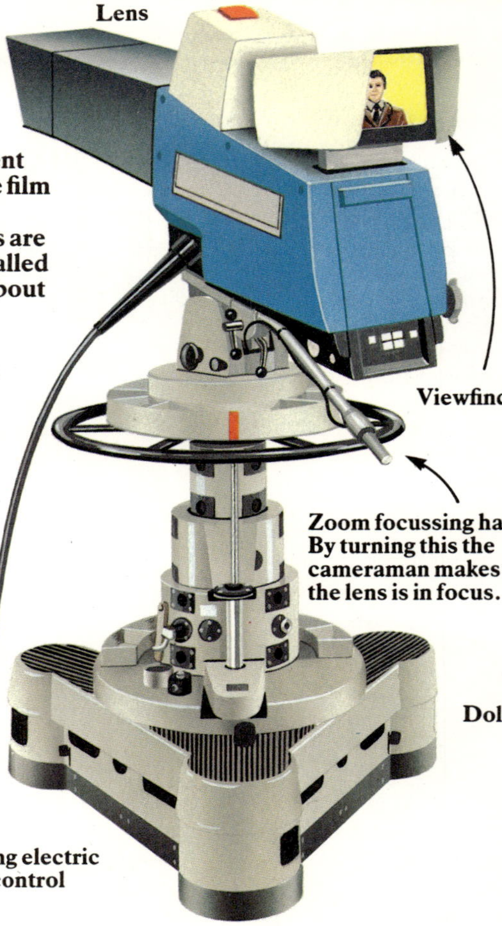

Lens
Viewfinder
Zoom focussing handle. By turning this the cameraman makes sure the lens is in focus.
Dolly
Cable taking electric signals to control room

Viewfinder

Cheaper cameras have an "optical" viewfinder, like those in still cameras. More expensive cameras have "electronic" viewfinders (mini TV screens) as in TV cameras.

Most video cameras have a zoom lens, which is either power-operated or hand-operated. Some also have a macro lens. This allows you to shoot subjects very close to the lens without going out of focus.

Handle for holding camera

Video cameras can be divided into three main groups. There are those designed for home use, like the one above, with which you can make your own programmes, record them on video cassettes and show them on your TV. There are video cameras designed for use in industry and education, which are slightly more complicated and produce better pictures. There are also video cameras designed to produce pictures to be broadcast, which are very similar to traditional TV cameras but much lighter and more manoeuvrable.

Camera tubes

The part of a camera that changes light into an electric current is called the tube. In a black and white camera there is one tube; in colour cameras there are usually three.

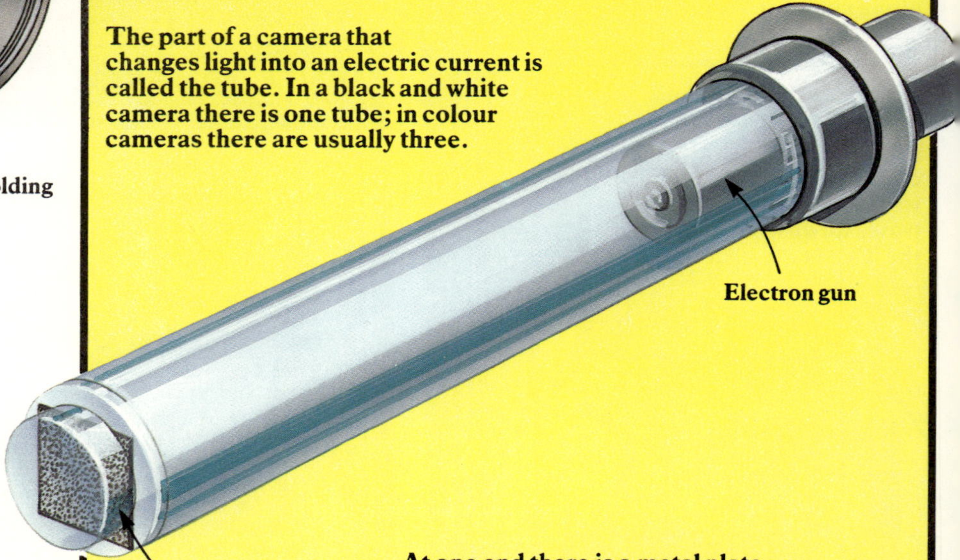

Electron gun
Target plate

At one end there is a metal plate called the target plate. At the other end is a device called an electron gun. You can see how the tube works on the opposite page.

Turning light into electric messages

To turn a scene into a picture on a TV screen, a TV camera has to convert the entire scene into a series of electric messages. It does this by turning light from the scene into electricity. TV sets turn the electricity back into pictures. It is easiest to understand how a camera works by first looking at how it converts scenes into black and white pictures.

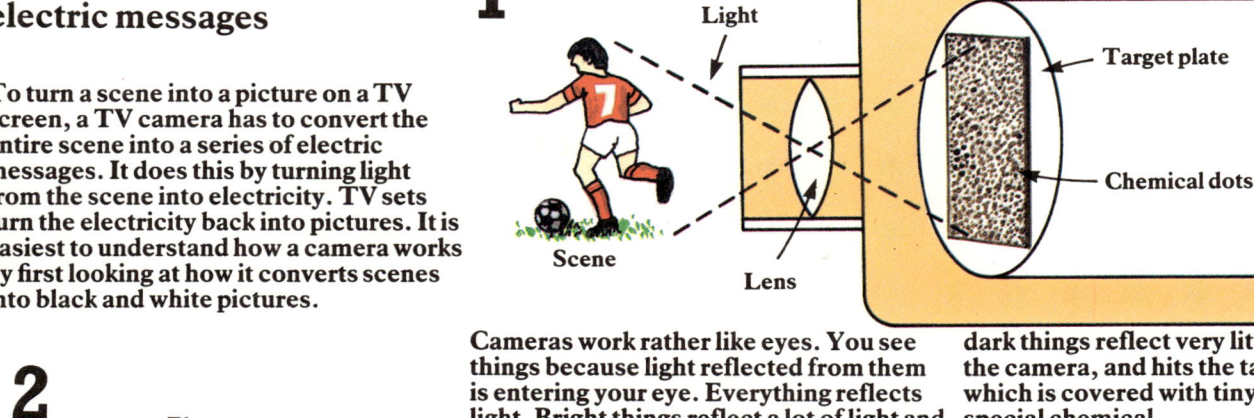

1

Cameras work rather like eyes. You see things because light reflected from them is entering your eye. Everything reflects light. Bright things reflect a lot of light and dark things reflect very little. Light enters the camera, and hits the target plate, which is covered with tiny dots of a special chemical.

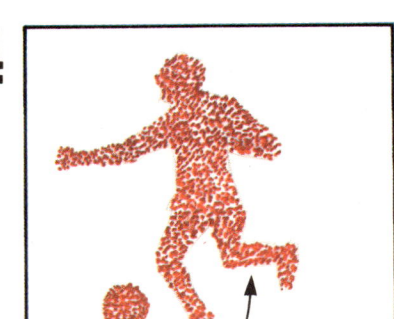

2

The chemical dots on the target plate are sensitive to light. When light hits them they react by releasing electrons (part of the atoms of which everything in the world is made). All atoms have a nucleus in the middle with electrons spinning round it. Each nucleus and each electron carries a tiny bit of electricity called a charge. A nucleus has a positive charge and an electron has a negative charge. Sometimes electrons split away from the nucleus leaving the atom with a positive charge.

3

When light falls on the chemical dots the electrons they release flow in the direction of the target plate. The size of the flow depends on the brightness of the light falling on each dot. The flow affects the atoms in the target plate and makes electrons escape down the tube.

4

Since electrons carry a negative charge, this leaves a pattern of positive charges on the back of the target plate. This now contains all the information necessary to reproduce the scene on a TV screen, but the information still needs to be carried out of the camera.

Taking the messages out of the camera

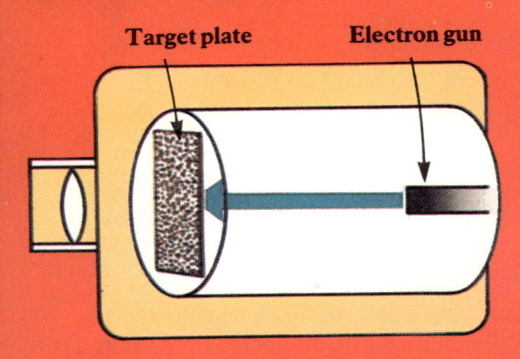

The electron gun is the device which interprets the information on the target plate and sends it out of the camera. It "reads" or scans the target plate by shooting a rapid stream of electrons down the tube towards it.

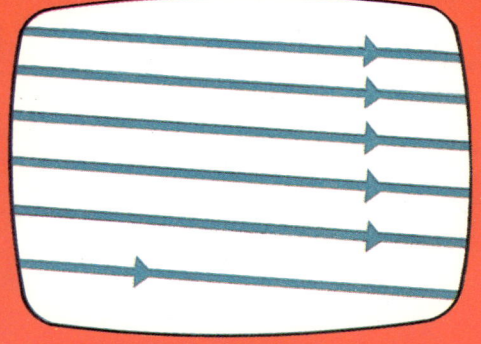

Lines show path of electron gun.

This diagram shows the path in which the stream of electrons moves over the target plate. It moves from left to right and from top to bottom in rather the same way as your eyes read the words on a page. It actually takes 625 lines (525 in the system used in America and Japan) to read the target plate from top to bottom and this only takes 1/25th of a second each time.

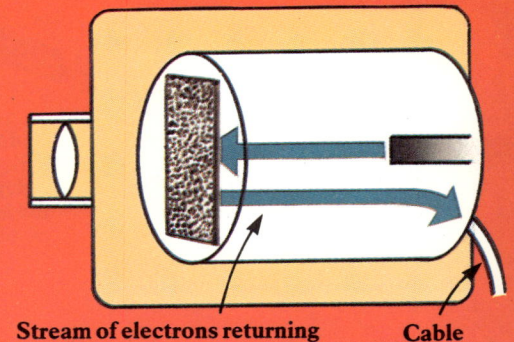

After hitting the target plate the stream of electrons returns down the tube. But where light from the original scene has made electrons escape from the plate, electrons from the stream replace them. This makes the returning stream vary in strength according to the brightness of the original scene. The stream of electrons passes out of the camera down a cable.

39

Colour cameras

Colour TV and video cameras work like black and white cameras, but they translate colours, as well as brightness, into electrical messages. To do this they have three camera tubes instead of one. Light entering the camera is split up into three key colours – red, green and blue – known as the primary colours of light. By mixing them together in various combinations and proportions you can produce any other colour, or even white. The three separate streams of light are directed on to the three separate camera tubes, which turn them into electric signals.

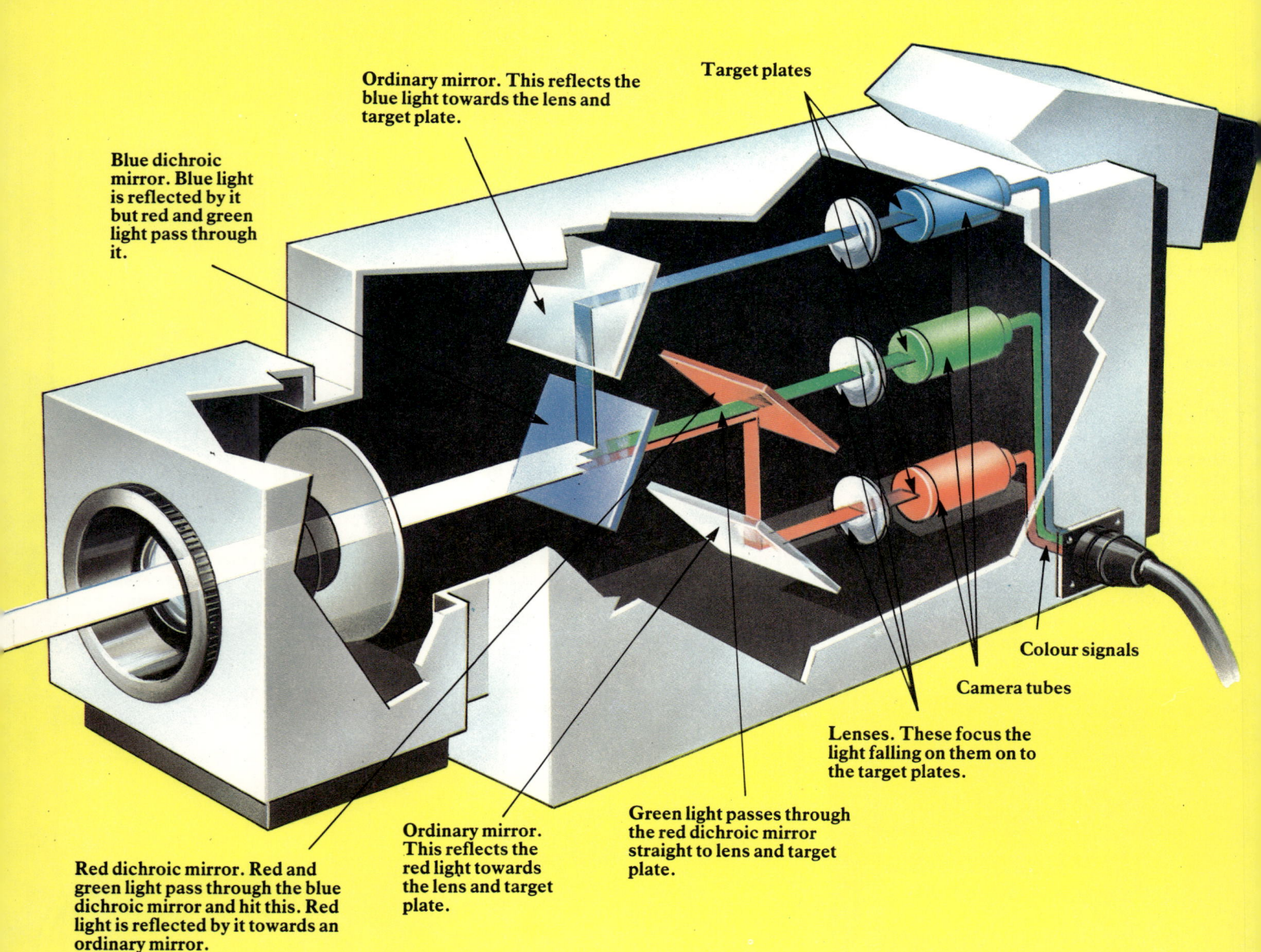

Ordinary mirror. This reflects the blue light towards the lens and target plate.

Target plates

Blue dichroic mirror. Blue light is reflected by it but red and green light pass through it.

Colour signals

Camera tubes

Lenses. These focus the light falling on them on to the target plates.

Green light passes through the red dichroic mirror straight to lens and target plate.

Ordinary mirror. This reflects the red light towards the lens and target plate.

Red dichroic mirror. Red and green light pass through the blue dichroic mirror and hit this. Red light is reflected by it towards an ordinary mirror.

Special "dichroic" mirrors split up all the light entering the camera into three separate streams of red, green and blue light. The three streams are each directed through three separate lenses on to three target plates. The target plates are then scanned by electron guns in exactly the same way as the target plate in a black and white camera is scanned. This produces three streams of electrical information, representing the brightness of each of the colours at each point in the picture. The three streams are then combined together to form one signal.

This is called the "luminance" signal. It conveys only variations of light and shade and can be received by black and white, as well as colour TV sets. Colour information (two "chrominance" signals combined together) is superimposed on top of this basic signal. Colour sets can decode this information and use it to add colour to the luminance signal.

There are different systems for coding the colour information on to the basic black and white signal. Most parts of Western Europe use a system called PAL; the French system is called SECAM; and the American system is NTSC. Other parts of the world have adopted one of these three systems.

Some video cameras have only one tube. The target plate is coated with a pattern of three kinds of dot. Each of which responds only to one of the three primary colours.

40

Television sound

A microphone turns sound into electronic signals, rather as a camera turns light into electronic signals. The sound signals can be stored on tape with the picture signals or transmitted immediately through the air on radio waves.

Various different types of microphone are used in making TV programmes. The type used depends on the situation and the kind of sound to be picked up. Sometimes as many as 200 are used for just one programme.

▲ Hand-held microphones are often used outside, where interviewers need to be able to move around.

▲ For drama programmes microphones are usually fixed to extendible metal rods called booms. A boom operator adjusts the length of the boom and the angle of the microphone so that it is close to the actor who is speaking but cannot be seen in the picture.

▲ Musicians usually use microphones fixed to adjustable metal stands.

◄ Small microphones pinned to a piece of clothing are often used on chat shows.

Picking up the right sounds

One of the problems sound engineers have to cope with is how to pick up the sound they want without also picking up all sorts of sounds that they do not want. They do this by selecting and positioning microphones very carefully and by adjusting the distance from which they will pick up sound. Different microphones have different sensitivities to the sounds around them.

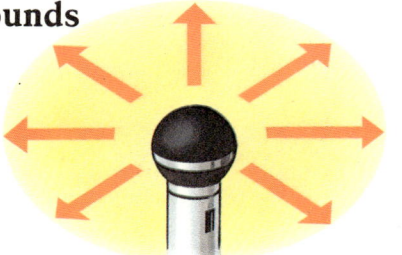

Some microphones pick up sounds equally from all directions. These are called omnidirectional or omnis for short.

A microphone that picks up sound only from the front and has a dead area behind it is called cardioid or directional.

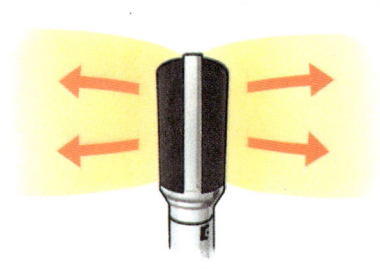

Some microphones pick up sound in a figure of eight area. These are bidirectional and are useful for two speakers facing one another.

How a microphone works

Sound of any kind is caused by something vibrating. What we hear as sounds are vibrations inside our ears. Vibrations travel through the air and make other objects vibrate. The stronger the vibrations the louder the sound.

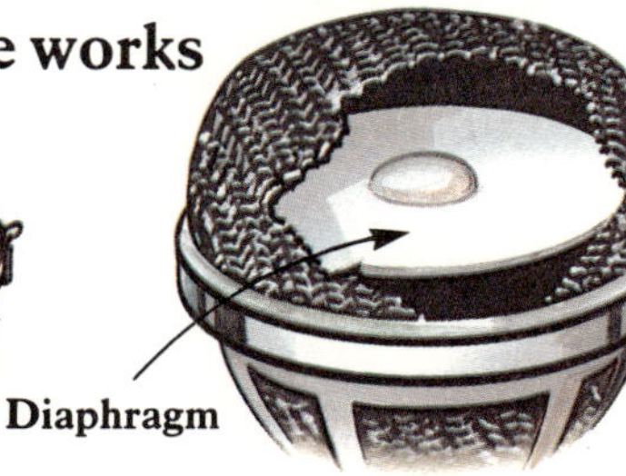

Inside a microphone is a very sensitive, metal plate called a diaphragm. Vibrations in the air, caused by sound, make it vibrate as well. The louder the sound the further it moves; the higher the sound the faster it moves.

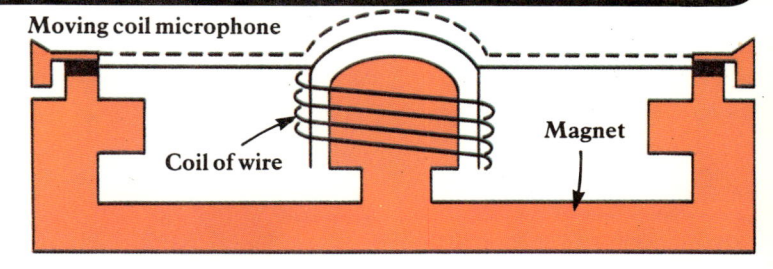

Attached to the diaphragm is a device which turns its vibrations into electric signals. The devices for doing this give different microphones their names – "moving coil", "ribbon", "capacitor".

In a moving coil microphone there is a magnet and a thin coil of wire attached to the diaphragm. When a wire moves near a magnet an electric current begins to flow. This current becomes the sound signal.

In a TV studio

Most TV programmes are made inside TV studios. A television station or centre usually contains several studios of different sizes. Attached to each studio is a set of three control rooms: the sound control room, the lighting and vision control room and the production control room. A central control room coordinates all the programmes and links material broadcast by the station.

Television screens called monitors show people in the studio the picture that will be transmitted.

The angle of the lights and the position of the shutters or "barn doors" in front of them, are adjusted by a long pole from the studio floor.

The cameramen wear headphones (known as cans) so they can hear instructions from the director, who sits in the production control room.

One camera is mounted on an electrically operated crane. This is used to take high angle and moving shots.

The floor assistant makes sure that the performers are in the right place at the right time.

The red line shows where the control room walls have been cut away, so that you can see the studio as well.

The stage manager checks that everything is in the right place.

Audience

Vision mixer

SOUND CONTROL ROOM

In the sound control room, sound engineers check the quality of the sound and make sure that the microphones are not picking up any unwanted noises. They adjust the volume and tone and add music, laughter and special sound effects, if they are needed.

PRODUCTION CONTROL ROOM

This is the production control room. The director sits here with the producer's assistant (PA), the vision mixer and the technical manager. The director and the PA have microphones in front of them. When they talk into them they can be heard by everyone

The lights hang from a grid on the ceiling. Each one has a motor and can be moved up and down by touching a switch on the studio floor.

Microphone

Actors

Sound boom

TV stations vary a great deal in size. Some of them are just big rooms, others are more like cinemas or theatres with tiered seats for a studio audience. In this one an episode of a sci-fi series is being made.

There are several different sets in the studio. Each one has the background and props for a different part of the action.

There are also several cameras. Each one shoots the scene from a different angle. The electronic signals from the cameras pass down cables through the vision control room to the production room, where the director chooses which picture will actually be transmitted. The sound signals pass from the microphones in the studio through the sound control room to the production control room.

For drama programmes there would not be an audience in the studio. Audiences are only allowed in for certain programmes such as comedies and chat shows.

A small light on top of each camera shows which camera is taking the picture that will actually be transmitted.

The floor manager is the director's contact in the studio. He receives instructions through his headphones and makes sure that everything runs smoothly.

Director   PA   Technical manager

wearing headphones in the studio and the people in the sound and vision rooms. Pictures from the cameras appear on the TV screens in front of them. The director gives directions to the cameramen, the floor manager and the vision mixer. To find out more about the control room see pages 44 and 45.

VISION CONTROL ROOM

All vision signals from the cameras go through this room to the production control room. The vision controller checks the technical quality and colour balance of the picture. Also from here the lighting supervisor controls the position and brightness of the lights.

43

In the control room

This is the production control room of a news programme. It is being broadcast live so everyone has to think very quickly to make it all run smoothly without mistakes. The director looks at the pictures from various sources, which appear on the monitors (screens) and decides which one should be broadcast at any moment and when to switch to another one. He gives instructions to the vision mixer, who operates the knobs and switches to achieve what the director wants. The director can also talk to the cameramen and floor manager through their headphones. If he wants to talk to the newsreader he has to press a special key. This is so that the newsreader is not distracted by a stream of instructions to other people.

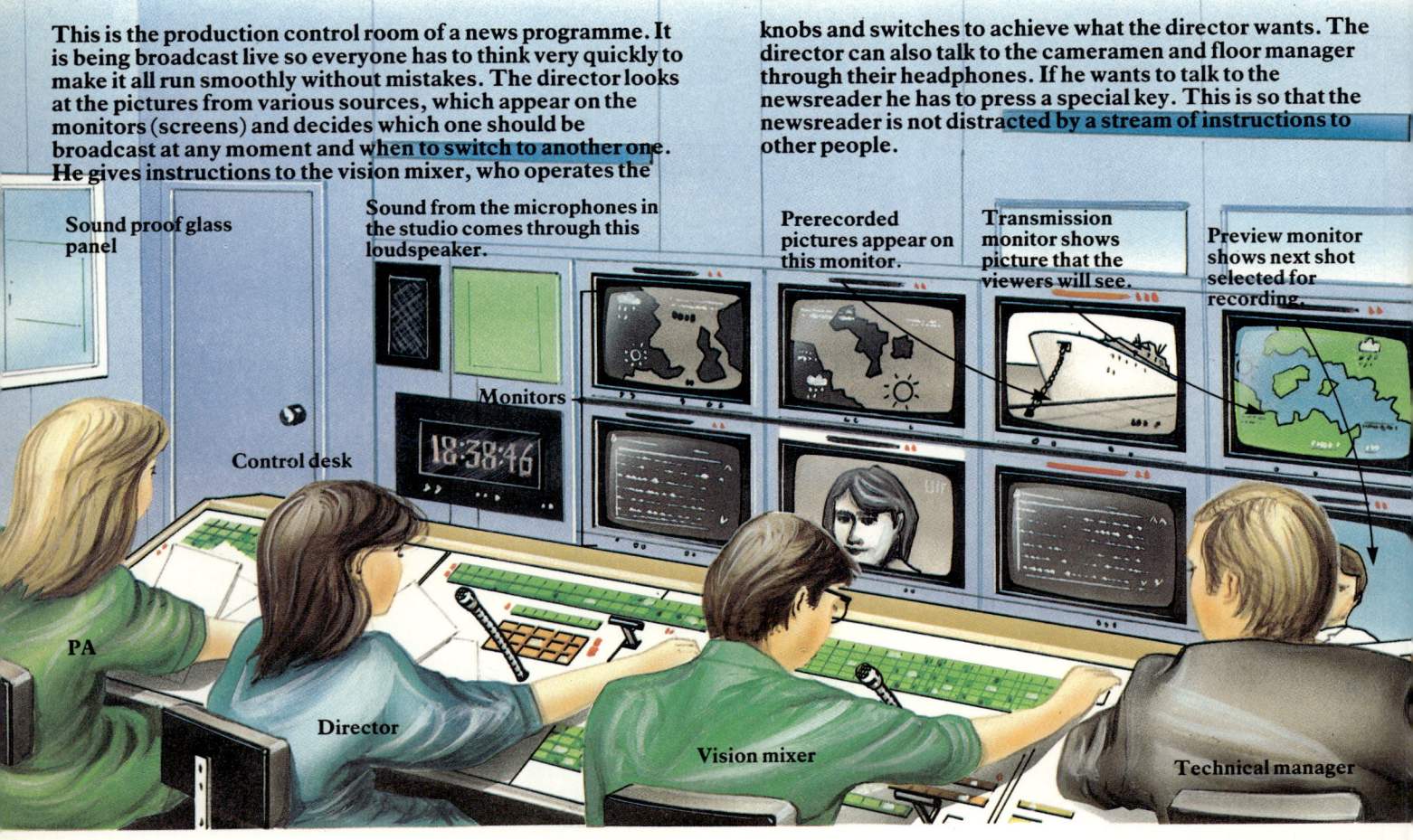

Sound proof glass panel

Sound from the microphones in the studio comes through this loudspeaker.

Prerecorded pictures appear on this monitor.

Transmission monitor shows picture that the viewers will see.

Preview monitor shows next shot selected for recording.

Monitors

Control desk

PA

Director

Vision mixer

Technical manager

Reading the news

The newsreader sits in a small studio near the production control room. He wears an earpiece through which he can hear instructions from the director.

A roll of paper with his words printed on it is fed across a table and past the lens of a small camera mounted above it. This machine is called a teleprompt. The picture taken by the teleprompt is fed to a small screen mounted at right angles below the camera lens, so that the newsreader can read his words while appearing to look directly at the camera. To the camera lens the words are invisible.

Any words, such as captions, titles or credits, that are to appear on the screen during the course of the programme are put on to an easel or loaded into a caption roller and placed in front of one of the cameras ready to be rolled past the lens. Machines called character generators* are, however, replacing caption rollers in many studios.

Teleprompter

Floor manager. He gives the newsreader his cues.

Caption roller

Monitor showing picture that is being transmitted.

Newsreader

*To find out more about these machines see page 57

The control desk

This is a close-up of the control panel from the production control room. The rows of buttons are known as buses. In each row across there is one button for each of the cameras and for each of the other sources, such as video tape machines, from which pictures are available.

If the vision mixer is cutting straight from one picture to another, he only needs to use one row of buttons. When he presses a button, it lights up showing that the picture from that camera is passing out of the desk to be transmitted.

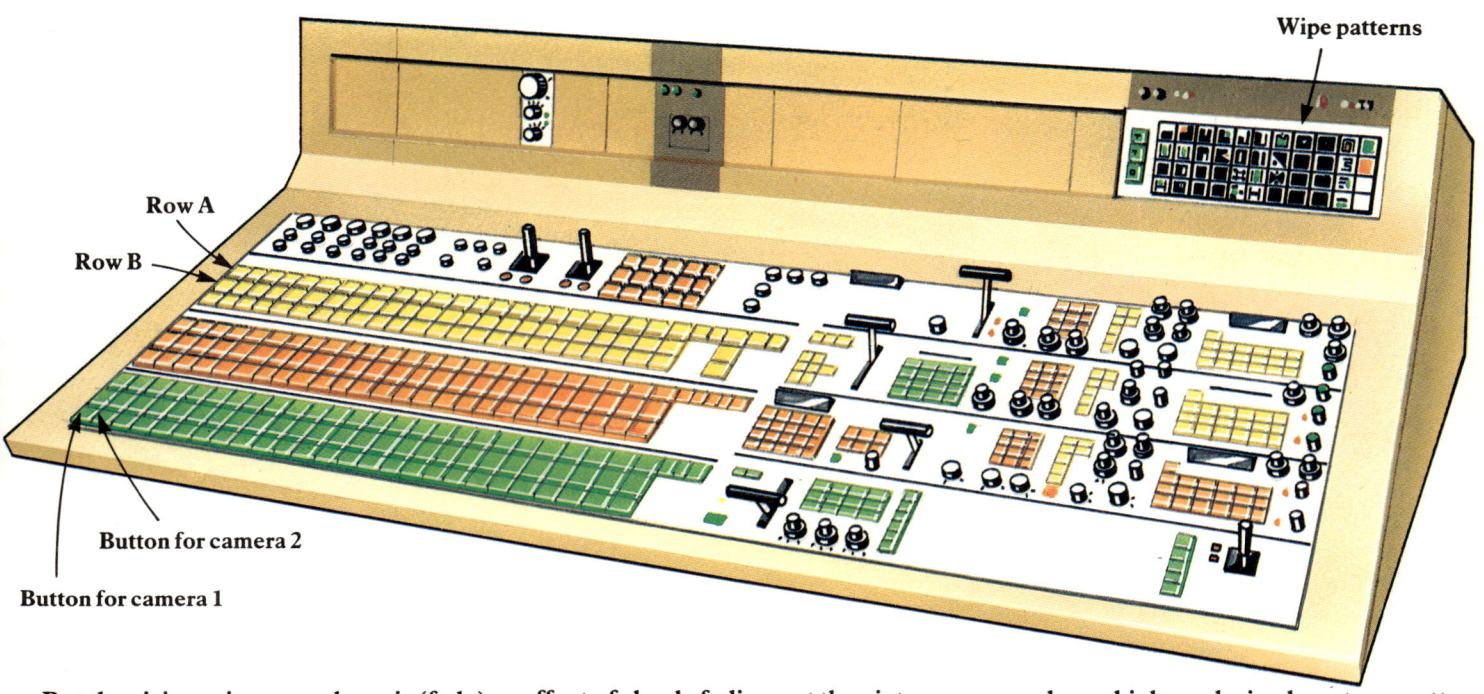

Row A
Row B
Button for camera 2
Button for camera 1
Wipe patterns

But the vision mixer can also mix (fade) one picture gradually into another. If he has selected camera 1 on row A and he wants to mix to camera 2, he presses the 2 button on row B and then moves the lever at the right slowly down. This has the effect of slowly fading out the picture from row A and slowly bringing in the picture from row B.

The change from one picture to another can also be done by a technique called wiping. This is when a moving edge, which can be in almost any pattern, leads the new picture across the screen, appearing to wipe the old one away. At the top of the panel a display of buttons shows some of the wipe patterns the vision mixer can select.

Cuts, mixes and wipes

Here you can see what cuts, mixes and wipes look like on the screen. These are the simplest ways of changing from one shot to another.

Cut   Mix   Wipe

45

Transmitting TV pictures

Most people receive the pictures on their TV sets through the air. Electric signals from cameras and microphones are sent along cables from the TV station to a transmitting station. There they are combined with radio waves and then sent out through the air to be received by aerials connected to TV sets in people's homes.

Sometimes underground cables are used to carry the signals. This happens particularly in places where there are special problems involved in transmitting signals through the air. To transmit programmes over very large distances, satellites are used. The use of both underground cables and satellites for transmission is increasing all the time.

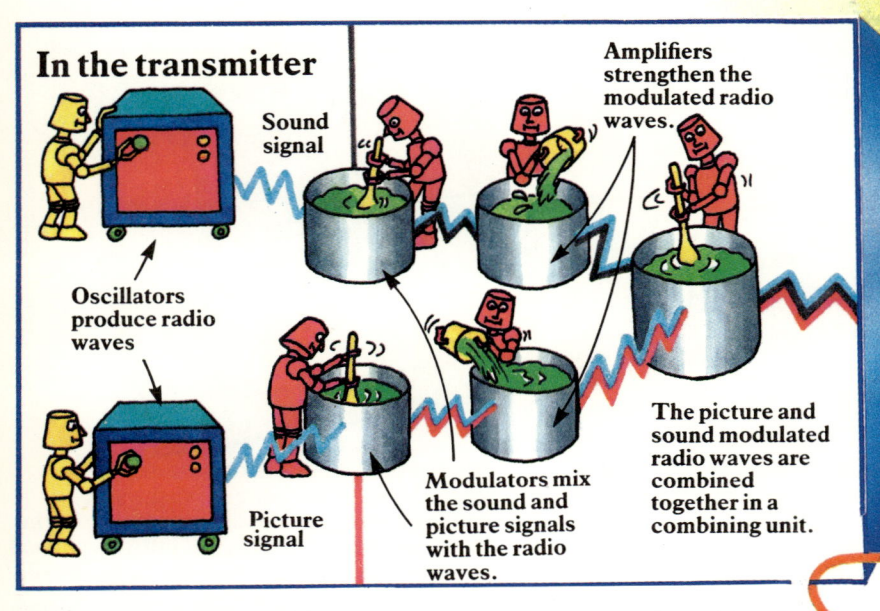

Radio waves go out in all directions, rather like light from a lighthouse.

In the transmitter

- Oscillators produce radio waves
- Sound signal
- Picture signal
- Modulators mix the sound and picture signals with the radio waves.
- Amplifiers strengthen the modulated radio waves.
- The picture and sound modulated radio waves are combined together in a combining unit.

In the transmitter there are two "oscillators". These make two radio waves – one to carry the picture information and one to carry the sound information. The two radio waves are sent to two "modulators" where they are mixed with (or modulated by) the sound and vision signals. The modulated radio waves are then amplified (strengthened by amplifiers), combined together and fed to the transmitting aerials.

◄ The transmitting aerial sends out the radio waves in all directions. The type of radio waves used to carry signals are made very much weaker if they travel through solid objects, so transmitting aerials are usually positioned high up on hills, or tall buildings, so that the signals do not hit any obstructions.

◄ The receiving aerial picks up the signal and sends it down a cable to the TV set.

Radio waves

Long wave (low frequency)

Short wave (high frequency)

Radio waves are described either by their wavelength (the distance between the top of one wave and the top of the next) or by their frequency (the number of waves per second). Long waves have low frequencies, short waves have high frequencies. TV signals are usually transmitted on very high frequency (VHF) waves, or ultra high frequency (UHF) waves.

All TV companies broadcasting in one particular area must use a different wavelength to transmit their signals. If they used the same ones or ones that were too close together, the information would get jumbled together and the TV sets would not be able to sort them out.

Transmitting by radio waves

Radio waves from TV transmitters cannot usually travel much more than about 80km. To broadcast over larger distances TV companies need to have several transmitters to pick up the signal from each other and pass it on. Where there are mountains, which would obstruct the signal, it must be carried over, either by a transmitter on top or by a land line. People living near mountains sometimes receive the same signal twice – once direct from the transmitter and once after it has bounced off the mountain and back to their aerial. The result is "ghosting", a faint shadow picture to the side of the main picture. To avoid this problem a local relay station is sometimes built. It picks up the signal from the main transmitter and broadcasts it on a different wavelength. People having problems receiving the main signal can tune their sets to a different frequency.

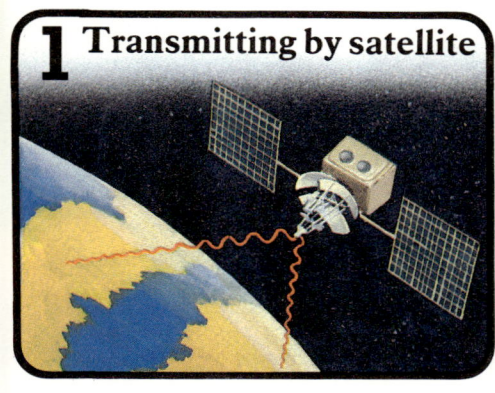

1 Transmitting by satellite

The best way of broadcasting over very large distances is to use a satellite circling several thousand kilometres above the earth's surface. Radio waves carrying TV signals are beamed up to it by a very powerful transmitter. When the satellite receives the signals it beams them back over a large area of the earth's surface.

2

The signal beamed back to earth by a satellite has to be picked up by a dish aerial. It can then be sent to viewers either by cable or by radio waves. As satellites capable of producing more and more powerful signals are developed, the dishes needed to receive the signals are becoming smaller and smaller.

3

The latest type of satellites are called Direct Broadcast Satellites. They give out such strong signals that they can be received by a dish aerial less than a metre in diameter – small enough to fit on the roof of a house, or in a small garden. The signal from the satellite can therefore go direct to individual homes.

1 Transmitting by cable

TV programmes are often sent from one country to another by underground cable and this system is also one solution for bad reception. A cable company puts up a high aerial and runs cables from it to customers who pay each month for the service.

The cables can carry a large number of different channels, but in most places only a few programmes are broadcast at any one time. So cable companies started offering programmes of their own. Some of them charge their customers more for each extra channel or certain special programmes.

2 Scrambled picture

In early pay TV systems viewers had to put coins in a slot. In the latest systems central computers record which channels customers have paid for. Pictures are sent out "scrambled". A box beside each TV set "descrambles" the picture if the computer approves it.

3

The latest development in cable TV systems is the fibre optic cable. The signals are carried by a laser beam – a very powerful beam of light. This is passed down a thread of glass, no thicker than an ordinary light flex. These cables can carry several hundred channels at a time.

TV sets

The job of a TV set is to turn the electronic signals created by TV cameras and microphones back into pictures and sound. The aerial picks up the signals from the transmitting aerials and they then travel down a cable to the aerial socket at the back of the TV. When you switch on the set the sound and picture signals are separated from each other and from the carrier waves. The sound is sent to a loudspeaker. The picture signal is sent to the picture tube, which converts it into the picture you see.

1 Tricking the eyes

If you look closely at a TV screen, you can see that the picture is made up of lots of horizontal lines. In Europe and much of the rest of the world TV screens have 625 lines, in America and Japan they have 525.

2

If you look even closer you will see that each line is made up of a series of red, green and blue dots (some TVs have strips instead of dots) of varying brightness. Most screens have over a million of these dots. The three colours mix in your eyes to produce all the colours you see on the screen.

How the picture tube works

The picture tube in a TV set works like the camera tube* in reverse. When the picture signal has been separated from the sound signal it is split into three separate signals – one red, one blue and one green. The tube converts these signals back into coloured light. The type shown here is called a shadow mask tube. There are other kinds but this is the most common.

The screen is the front part of the picture tube. On the inside it is covered with tiny dots of a chemical called phosphor. Three different types of phosphor are used. One type glows red when hit by electrons, one glows blue and one glows green.

At the back of the picture tube are three electron guns that fire beams of electrons at the screen. The amount of electricity leaving the guns is controlled by the three picture signals which are fed to the picture tube. A strong beam makes the phosphor glow brightly and a weak beam makes it glow dimly. The beams move across the back of the screen line by line.

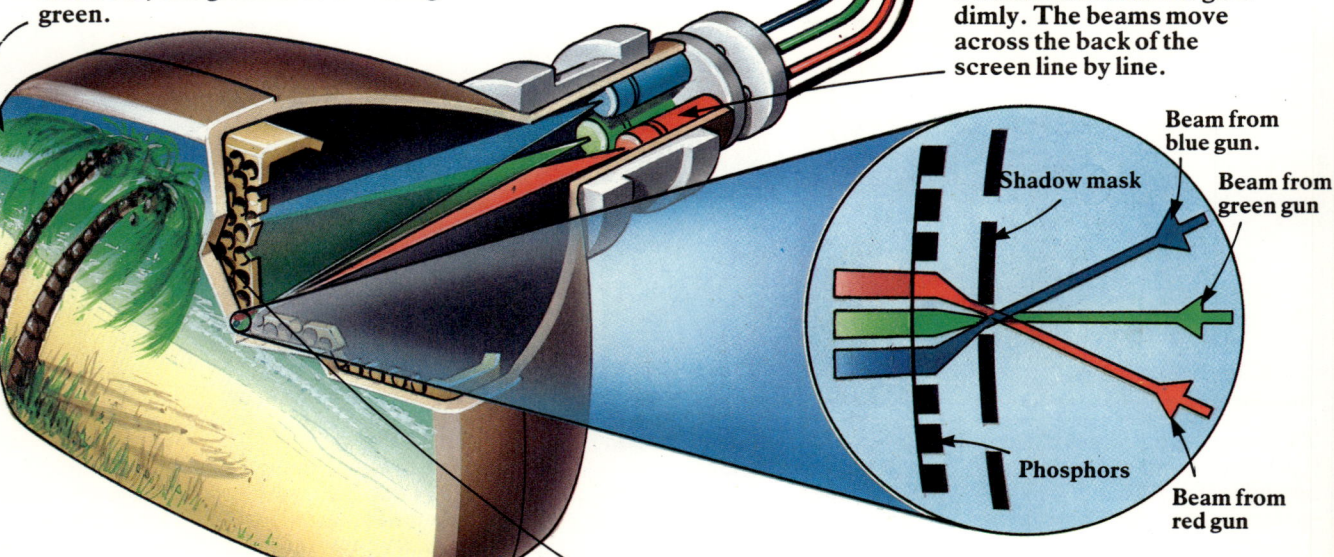

Beam from blue gun.
Shadow mask
Beam from green gun
Phosphors
Beam from red gun

Behind the screen is a metal plate called a shadow mask. It has thousands of tiny holes in it. There is one hole for every three dots of phosphor on the screen. The holes are positioned in such a way that each of the three electron beams can only strike the right type of phosphor.

*See page 38

3

The picture on the screen is actually created one dot at a time. Each dot in turn lights up and fades line by line down the screen. Any picture that the brain receives takes 1/10th of a second to fade away, but the dots light up and fade so quickly, that every dot on the screen appears to be lit all the time.

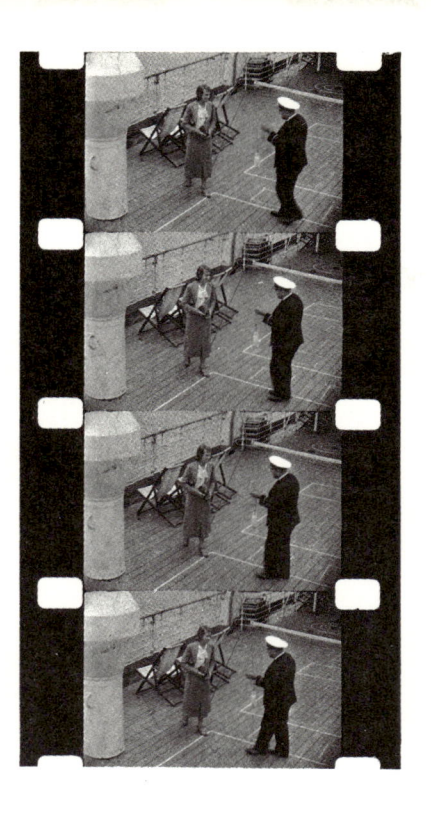

4

The slow-to-fade effect of human sight is called "persistence of vision". Cine films use this factor to create an impression of continuous movement. They are in fact a series of still photographs or "frames" taken in very quick succession – usually 24 frames per second. When they are projected at the same speed, our eyes see one continuous picture.

In TV systems that have 625 lines per frame TV frames are shown at the rate of 25 complete frames every second. This means that when films are shown on television they run fractionally faster than they would do in a cinema, but the difference is so small that you can hardly notice it.

Selecting a channel

Most TV sets offer a choice of programmes from several different broadcasting stations. The radio waves that carry the TV signals are divided into channels according to their wavelength and frequency.* Within each area, each station is given a different channel, so that they will all be using different frequencies to transmit their signals. There are gaps between each channel to make sure the signals do not interfere with each other.

When the radio waves enter the TV set they pass into a tuner. The tuner will only allow waves within a certain band of frequencies to pass into the rest of the set. But the tuner is adjustable and can be altered to allow a different band of frequencies in. You adjust the tuner by turning a knob or pressing buttons on the front of the TV set. When you first get the set or you want to pick up a new broadcasting station the tuner has to be adjusted so that when you press one of the buttons or turn the knob it allows in the band of frequencies on which that station is broadcasting in your area. Each knob can be preset to any channel.

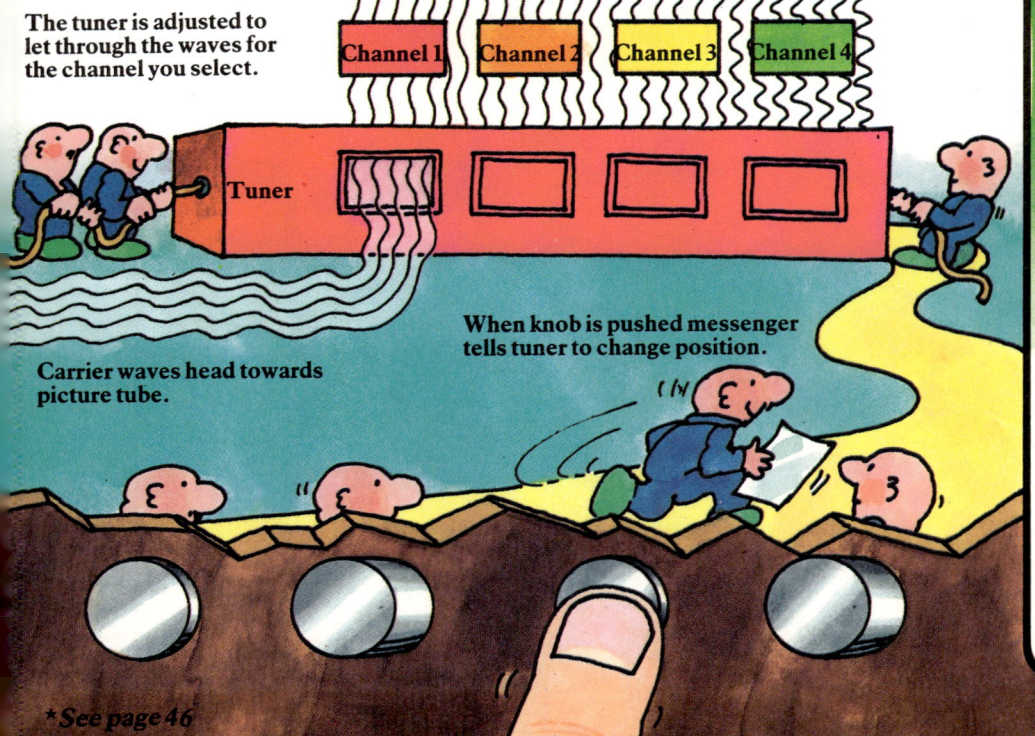

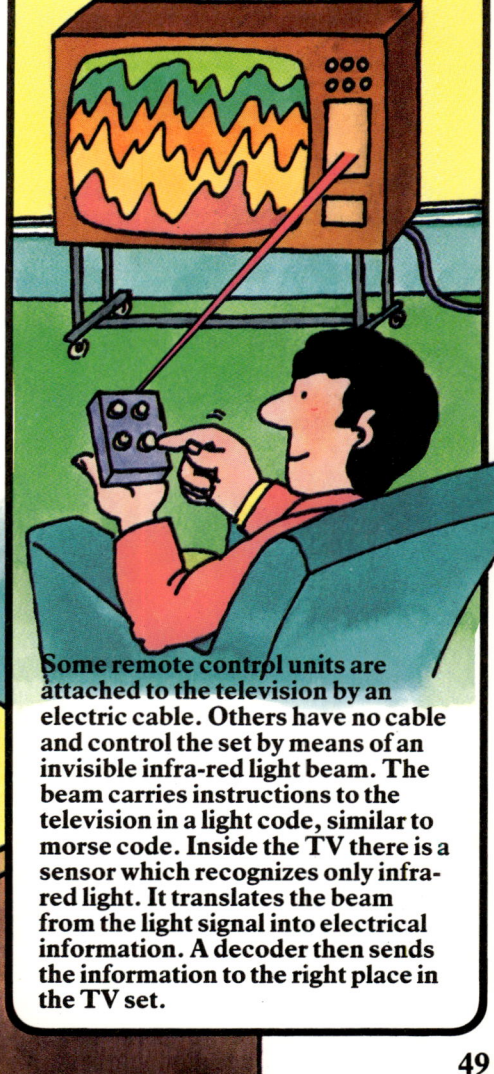

The tuner is adjusted to let through the waves for the channel you select.

Carrier waves head towards picture tube.

When knob is pushed messenger tells tuner to change position.

Remote controls

Some remote control units are attached to the television by an electric cable. Others have no cable and control the set by means of an invisible infra-red light beam. The beam carries instructions to the television in a light code, similar to morse code. Inside the TV there is a sensor which recognizes only infra-red light. It translates the beam from the light signal into electrical information. A decoder then sends the information to the right place in the TV set.

*See page 46

49

Outside the studios

Programme makers often have to go outside their TV studios for the material they need. This presents different problems from those that arise in studio productions and they have to work out what equipment they will need to solve them. Film equipment is lighter and more manoeuvrable than normal studio equipment and in some situations this is very important. But film does have to go through more processes before it can be broadcast. Events that need to be covered by several cameras and broadcast live or almost immediately are covered by "outside broadcast" units. This means taking along TV cameras and a travelling control room with a team of people to operate the equipment. The development of much lighter electronic video equipment is, however, beginning to bring about great changes in the way outside material is gathered.

The centre of operations in an outside ▶ broadcast is the mobile control room (MCR). This is a large van which contains the equipment normally found in the production, sound and vision control rooms linked to a TV studio. All the pictures from the TV cameras are sent back here and the ones selected for broadcasting are then sent back to the TV station by radio waves or by cables.

Outside broadcasts

Most sports programmes and other events not staged specially for the TV cameras are outside broadcasts (usually referred to as "Oh bees") and parts of other programmes are made in this way too. The location need not necessarily be outdoors. O.B.'s are often broadcast from theatres, concert halls and other indoor locations.

Where land lines (cables) exist, signals are sent back to the TV station through them. If there are none available, the signals are sent by radio waves from a portable transmitter set up next to the van. The transmitting and receiving aerials are usually dish-shaped.

Commentators usually have a position overlooking the whole scene, but they also have monitors showing the pictures from each camera. They base their commentary partly on what they can actually see and partly on the monitors, so that can fit their commentary to what the viewer can see.

Production control area
Sound engineer
Vision engineer
Gun microphone
Camera 1
Camera and microphone cables plug in here.

1 News gathering

Speed is vitally important when gathering news stories. Until quite recently most news pictures were obtained on film. Small two-man film crews can get to scenes of action very quickly, taking all their equipment in one van or car.

2

Once the film has been shot it has to be taken back to the studio as fast as possible. A motorbike messenger is often used. In the TV station it has to be developed, edited and put through a telecine machine before it is ready to be transmitted.

3

Now that smaller and lighter cameras and video tape equipment have been developed, a two-man TV camera team can also get news pictures without taking a lot of cumbersome equipment with them. This is called electronic news gathering (ENG).

As in studio broadcasts, several cameras are used in an outside broadcast. They are often the same type as studio cameras but some O.B. units now use smaller, lighter ones. They may be mounted at ground level or on top of vehicles, buildings or scaffolding platforms.

Camera 2

Camera 3

Outdoors it is often impossible to place a microphone near the sound source, so a gun microphone is used. This can pick up sound from quite a distance away if it comes from the direction in which it is pointed. A windshield is often fitted round the microphone to reduce noise caused by the wind.

Microwave radio links

The type of radio waves used to carry signals from the mobile control room to the TV station, when there is no cable link available, are the very short ones called microwaves. They can also be used to carry the signals from the cameras to the control room, in cases where a cable link would prove too awkward. Microwaves are used so that these signals do not interfere with signals being broadcast to people's homes. They cannot travel through objects so they often have to be "bounced" to a series of different aerials before reaching the TV station.

4

The tape can be taken back to the TV station by a messenger, but it is also possible to send the pictures by land line or by microwave radio link. The crew often travel in vans with a microwave transmitter on top. At the station the pictures can be transmitted instantly.

5

Some of the latest cameras now have a compact video cassette recorder built into the main body of the camera. This makes it possible for one person to record pictures and sound and send them back to the station.

Filming on location

When programme makers decide to use film cameras for shots outside the studios, this is called filming on location. Many drama programmes contain filmed sequences. The location may not be outdoors. Often it is very hard to make shots in a studio look realistic and it may be cheaper to use an existing building to get the right kind of atmosphere and detail.

Cameraman
Assistant cameraman
Sound recordist
PA
Director

One camera is used and each shot is carefully planned and prepared in advance. The sound and pictures are recorded separately. The clapperboard is used at the beginning of each shot, so that the sound and the picture can be exactly matched, or synchronized, when they are put together.

Electronic field production

The newer, lightweight electronic equipment is now beginning to be used for some drama programmes where film would previously have been used. This way of working is rather similar to the procedure in making home videos.

Recording and editing

Most programmes are recorded and edited before being transmitted. Sometimes film is used but more often they are recorded on video tape.

How an electronic signal is recorded on tape

To record a sound or vision signal on tape, the electronic signal must first be converted into a magnetic signal. This is done by a recording head.

The unrecorded tape is fed to the recording head. the tape is made of plastic coated with a layer of tiny particles of iron oxide. As it passes across the recording head the particles of iron oxide are magnetized by the iron block. Their magnetic strength varies with the magnetic strength of the block at the moment they pass across it. In this way the signal is stored on the tape in the form of a magnetic pattern.

For sound recording the tape width is ¼ or ⅛ of an inch. In video recording there is much more information to store. In TV broadcasting 2 inch or 1 inch tape is used. Tapes for use in industry are ¾ inch and cassettes for home use are ½ an inch. There are also now ¼ inch video tapes for use in some portable video recorders.

When the tape is rewound and played back the recording head "reads" the magnetic pattern and converts it back into an electronic signal.

▲ Video tape recorders (VTRs) are used for both recording and editing tape. They are the professional version of the domestic video cassette recorder (VCR). The tape is on large reels instead of cassettes.

A recording head consists of a coil of wire wrapped around a small block of iron. When the signal reaches the wire the iron block becomes magnetized. Its magnetic strength varies with the strength of the electronic signal.

Moving heads

In sound recording the sound is recorded in straight lines down the length of the tape and the recording head does not move as the tape moves past it. For video recording, ways had to be developed of packing a much greater volume of information on to the tape in a more economical fashion. If the sound recording system were used for video recording you would need more than 100km of tape to record a one hour programme.

To overcome this problem the recording heads on video tape recorders move as the tape moves past them. By laying the information across the tape they fit more of it into the same amount of space.

In VCRs and the newer VTRs used in broadcasting a "helical" (spiral) system of moving recording heads is used. Two, sometimes four, heads are mounted on a wheel inside a drum. The tape wraps round the drum in a spiral path and, as it slides across the drum, the wheel with the heads rotates in the opposite direction. The information is laid on the tape in long, slanting tracks. Sound is transferred to a strip down one edge of the tape by a separate sound recording head.

Editing

Editing is the process of assembling a series of different scenes or "shots" into one continuous programme. It involves cutting out certain shots, or parts of them, and changing the sequence of others. The editor may need to cut the overall length of a programme, or to cut out mistakes and retakes. The impression each shot makes on the viewer can be completely altered by cutting out a few frames.

Electronic editing

Video tape editing is done by dubbing (copying) material from one tape on to another. At least two machines are used – one to playback the first tape and one to record the selected pieces from the first tape on to the second tape.

Editing requires split second timing. In TV stations editing suites, like the one shown here, usually use computers to control the process. This helps to ensure that the joins are made in the exact place chosen by the editor and director.

You cannot edit videotape by cutting it up and joining together the pieces you want. This is partly because you cannot see the images on the tape so it would be very difficult to know exactly where to cut, but also because cutting tape creates enormous picture disturbance and joins can damage the recording heads.

Film editing

In film editing you can actually see the image on each frame of the film. The editor cuts the film and joins the selected pieces together with transparent tape. To find shots quickly all the bits of film are carefully filed in cans. The room where film editing is done is called the cutting room.

Use this to see what you are recording.

Check what you have recorded on this.

VCR 1

VCR 2

Editing at home

To edit tapes at home you need two video cassette recorders connected to each other by the video out/in and audio out/in sockets. It helps if you also use two TV sets, but you could make do with one.

First view your tape and decide which sections you want to use and in what order. Use the tape counter to help you remember. Then record the sections in the selected order by playing them on VCR 1 and putting VCR 2 on record.

The aim is to get the joins as invisible as possible. If VCR 2 has an edit start button it will help to cut down picture distortion. It also helps if tape 1 is already running when you start to record on tape 2. Start tape 1 a little before the section you want and use the pause control on VCR 2 to bring in tape 2 at exactly the right moment. The quality of the picture will be a little worse on tape 2 than on the original.

Special electronic effects

The picture that appears on your TV screen is not always exactly like the live scene originally taken by the camera. The camera can only take what it sees, but in the control room or editing suite the image from one camera can be electronically combined with other images. Combinations of different images in one picture can be used to create very strange and magical effects, but they can also produce the kind of results we have become used to seeing on news and current affairs programmes. These two pages explain how some of the effects are produced. When you are watching TV see if you can tell which technique has been used.

1 Double shots

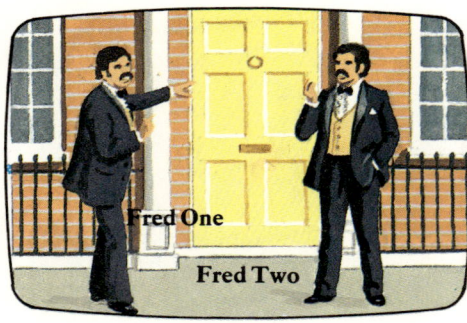

One of the effects that modern recording methods can achieve is a double shot. An actor (Fred One) can appear on the screen with himself (Fred Two) as though he were two separate people. He can even carry on a conversation with himself.

2

First the actor plays the part of Fred One. He has to be careful to keep to the left of the set and to leave pauses long enough for Fred Two's replies. The scene is recorded on video tape A and the tape is then rewound.

Chromakey

Chromakey is the most common technique used for combining two pictures. It can be used to show a map or photo behind a newscaster or presenter, or to show more fantastic and magical scenes.

1

Chromakey works by cutting out all parts of a picture that are in one "key" colour. A bright blue is usually chosen as the key colour. In the picture above, the newsreader sits in front of a blue background. The camera's output is passed through an electronic switch. As each line of the scene reaches it, the switch checks whether it is blue or not. Wherever it finds a patch of blue, it rejects it and switches instead to another source. The programme director can choose any other source. Here he has chosen camera 2, which is focused on a map, but he could choose a picture from an outside broadcast or satellite relay from another part of the world. The newscaster must be careful not to wear anything blue, because if he does, part of his body will seem to disappear.

2

For a flying carpet scene, the carpet can be laid out on a floor which has been painted blue. The camera must look down on the scene so that all the background colour is blue. The background can come from film of the sky taken from an aeroplane.

3

The switch can be adjusted so that instead of rejecting the key colour, it lets through only the key colour. By using this technique the background can be made to remain, while figures, or other foreground shapes disappear, leaving only their outlines. The outlines can then be filled in by shots from another source.

3

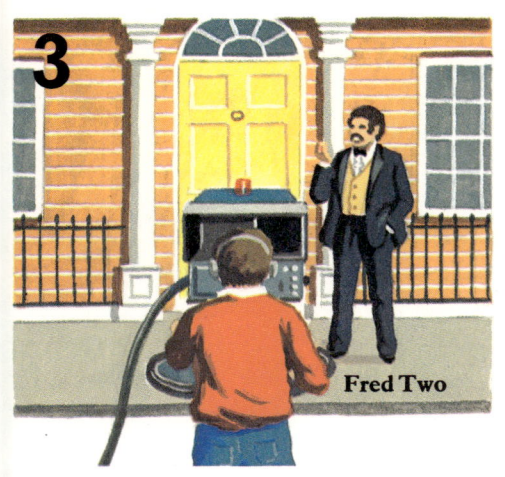

The camera then shoots the same actor, now playing the part of Fred Two. This time he is careful to keep to the right-hand side of the set and to leave pauses for Fred One's parts of the conversation. The background must exactly match the first shot.

4

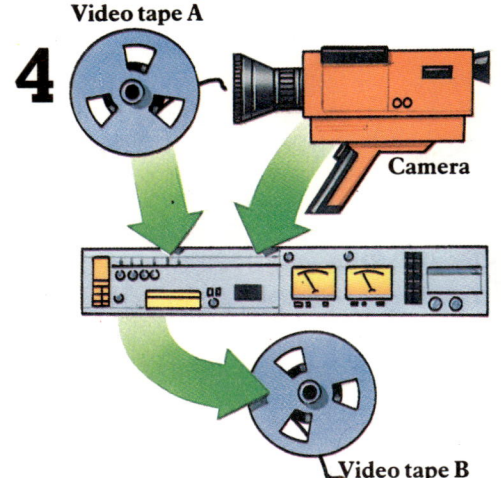

The signal from video tape A is sent to the mixing desk at the same time as the shot of Fred Two. In the desk the left side of the tape shot and the right side of the camera shot are taken and combined together and recorded in the final version on tape B.

4

Chromakey can also be used to distort the scale of things in a scene. Here a timid man is scared by a giant cat. To achieve this effect camera 1 shoots a close-up of the cat against a blue background. Camera 2 shoots the man from quite a distance away. Shot 2 is then fed into the blue parts of shot 1.

5

It is also possible to use chromakey to make someone disappear from the screen. The figure is keyed in from a second camera. You can then slowly mix from the composite shot to the background alone with the effect that the figure appears to vanish.

Painting by numbers

One simple electronic effect can imitate the appearance of a picture done by "painting by numbers" and give the picture on the screen a cartoon-like effect. This is done by using a switch, like the one used for chromakey, which can respond to a number of different levels of brightness in the picture. Each level of brightness can then be replaced by a different colour.

Double shot with chromakey

The techniques used to show two life-sized Freds can be combined with the chromakey scale-distorting effect so that Fred Two appears to be a pin-sized man, dancing in the hand of Fred One. On the second recording Fred Two is photographed from far away against a blue background and overlayed on a prerecorded image of Fred One. The cameraman has to be careful to line up the feet of Fred Two so that he appears to be in the hand of Fred One.

Digital effects

Some very spectacular effects can be achieved on the TV screen by changing the shape and size of one or more picture frames. This is done by feeding the signal from the TV camera through a special computer. The signal from the camera is electronic and has to be converted into a number code before it can be processed by the computer. A signal based on numbers is called a digital signal, so special effects created by this method are called digital effects.

Here are some of the effects you can see on television which are achieved by using digital signals. See how many of them you can spot when watching TV.

A digital effects machine

The computer which changes the size and shape of the picture is known as a digital effects device. It is a large box inside which there are several boards with silicon chips attached to them.

Signal from TV camera

ANALOGUE-DIGITAL CONVERTER

The analogue signal from the TV camera enters the machine here. The first thing that happens inside the machine is that it is converted into a digital signal. This is done by testing it at regular intervals thousands of times a second. The strength of the signal at each point tested is interpreted as a number.

Instructions from operator arrive here.

CENTRAL PROCESSING UNIT

Instructions from the operator are fed straight into this section, which decides mathematically how to use the information in the store to create the effect the operator wants. It has just received the instruction "reduce the size of this picture to one quarter its present size".

Analogue and digital signals

The normal signal used in television is an electric current generated by light falling on the target plate of a camera tube and is called an "analogue" signal. It is a continuous signal and its strength varies in exact proportion to the strength of the light falling on the target plate.

Analogue signal

A digital signal is a series of electric pulses. The pulses form a number code. The strength of the light at any given point is represented by a number.

56

The operator uses a separate control panel to feed instructions into the box.

The control panel is attached to the box by a cable.

FRAME STORE

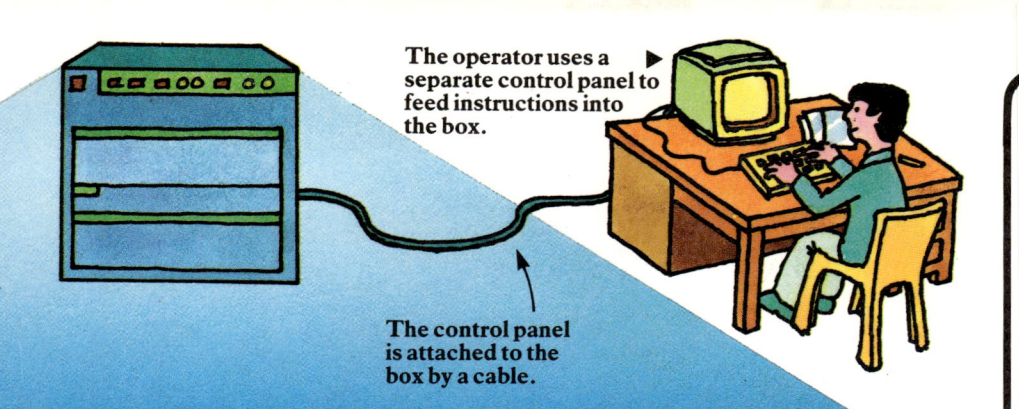

The numbers are carried in the form of electric pulses to the frame store. The frame store can only hold information about one complete picture frame at any one time. As soon as this information is taken out of the store and used, it starts filling up with information about another frame.

CLOCK

The clock is a vitally important part of the machine. All the stages described here take only thousandths of a second to happen, so the timing has to be extremely accurate.

Messages are sent to the frame store to tell it to send out one quarter of the numbers it has stored – each alternate number from each alternate line. These numbers are sent to the fourth department – the digital-analogue converter.

DIGITAL-ANALOGUE CONVERTER

Here the information about the quarter-size picture is translated from a digital signal back into an analogue signal, which can be carried by cables and transmitted to people's TV sets in the normal way.

Character generators

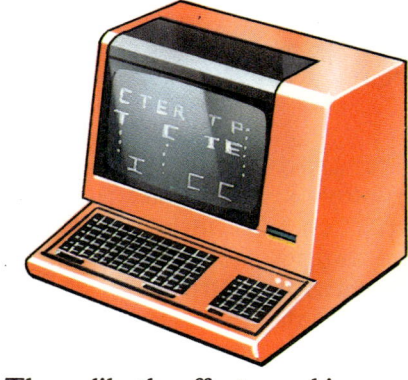

These, like the effects machine, are digital devices that allow letters, numbers and symbols to be typed into a keyboard and appear straight away on the TV screen. The information can be moved about the screen, flashed on and off, or rolled up to look like a roller caption.

Digital paint boxes

When you draw or write on the tablet of a digital paint box the result appears directly on the TV screen. The tablet responds to pressure by producing an electronic signal. The width and texture of the lines can be changed and you can fill in solid areas with colour.

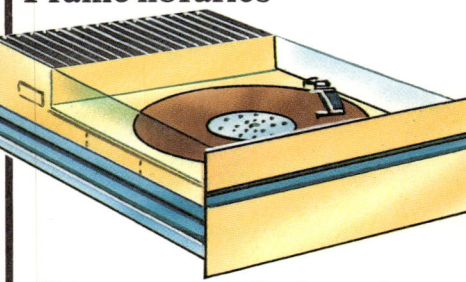

Frame libraries

Using a computer disc for storing information, it is possible to file away hundreds of still TV frames. This means that a whole library can be built up from pictures frozen and grabbed from the TV, or from slides. These can be retrieved instantly and slotted into a programme.

TV and computers

The rapid progress of computer technology over the last few years, and particularly the development of home computers, is turning the TV set into a multi-purpose home terminal, linking you into all kinds of information systems other than just broadcast programmes. On these two pages are some of the computer uses of TV which are already in operation and some which are possible and likely to be used before too long.

There are lots of different ways of putting information into a computer, and of getting results from it. One of the ways a computer can display its results is by showing words, diagrams or graphs on a screen called a visual display unit (VDU) or monitor. A TV screen can serve as a display screen for a computer. The information is sent to the TV in the same form as a broadcast signal. It is then sent to the electron gun, which transfers the information to the screen.

Home computers

There is now a wide range of small computers available for home use. They can be used for playing games, for storing information, for making calculations and for learning how to do things. Home computers are designed for use with an ordinary TV set, which acts as its display screen, though you can also buy specially designed monitor screens. The computer itself comes in the form of a keyboard unit which has letters and numbers, like a typewriter, and also a set of instruction keys. The keyboard connects to the TV by a lead which plugs into the aerial socket.

Some computers are capable of producing colour on the screen and some will also produce sound.

TV games

A TV games system consists of a small computer, called a console, and a set of hand controls – usually joysticks, paddles or pressure pads. The console plugs into the aerial socket of your TV. The information the console needs to play each different game is contained on a silicon chip. Each chip contains about ten variations of the same game.

There are two types of TV games systems. In the first the games chip is fixed permanently in the console. In the second type you buy the games chips separately. They come inside plastic cartridges, which slot into the console. The more expensive the game you buy, the more sophisticated the display becomes and the more control you have over what happens on the screen. As time goes by the gap between TV games systems and home computers is gradually closing. Some of the newer systems provide you with a keyboard to put in your own games programs as well as using preprogrammed chips.

Computer information banks

TV sets can already be used to receive information coming from a central computer information bank outside your home. They are likely to be used more and more for this in the future.

The information can be sent to the TV in two different ways. It can be transmitted along with the normal TV signal and picked up by the TV aerial, though the TV set has to be specially adapted to receive it. This system is called teletext. The other system, called videotex, or viewdata, sends the information as signals along telephone wires. They have to be decoded in a special unit, called a modem, before the TV can understand them.

Both systems display screenfuls of the latest information on all kinds of subjects, including news, sport, weather, travel, finance and business. It is like having an instantly updated newspaper or magazine.

Transmitting programs

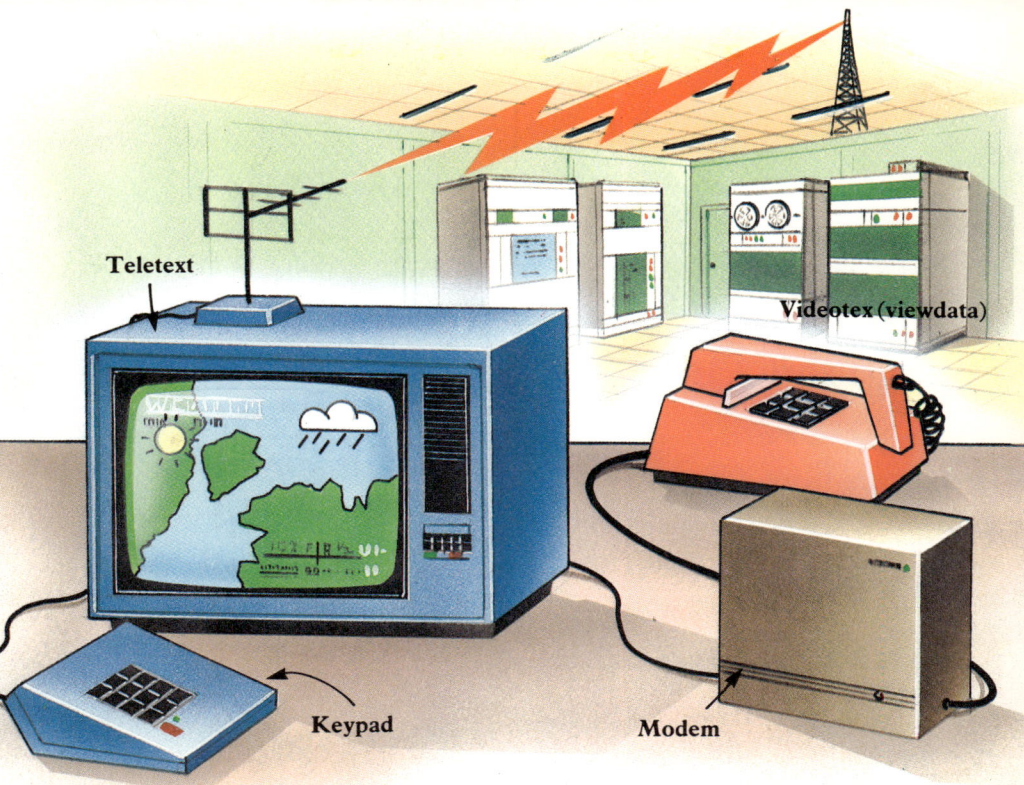

To make a computer do anything you have to feed in a list of instructions and information, called a program. Instead of transmitting pages of straightforward factual information, the central computer could transmit pages of computer programs for use on home computers. The program could even be stored for later use on an audio tape cassette or a floppy disc linked up to the computer.

Interactive TV

Various systems of "talking back" to the TV set have been tested in different parts of the world. Viewers are provided with a keyboard and second screen. Questions come up on the second screen and information passes from the keyboard, via the telephone lines to a central computer in the TV studio. This system can be used for testing audience reaction to programmes, taking opinion polls and playing games.

Teleshopping

In the future people may do much of their shopping by watching television. A TV in the home could be linked via a home computer to a computer in a shop. Goods from the shop would appear on the screen and you could use your computer to send orders to the shop's computer. It would give instructions to send you the goods and a bill.

Telebanking

There are already systems which allow you to call up your bank statement direct from the bank's information centre, if you have the right equipment. This could be extended to cover all your transactions with your bank. You could transfer funds and pay bills by typing instructions on a keyboard.

Teledoctor

Experiments have been made in diagnosing illnesses by using computers and the results have been surprisingly accurate. This could be developed so that if you felt ill you could call up a program on your screen, which would ask you a series of questions and then tell you what course of action to take.

Home video equipment

The development of electronic cameras and recorders which are small and cheap enough for people to have in their homes, is bringing about a revolution in home entertainment.

The central piece of equipment in the video revolution up to now is the video cassette recorder (VCR). The number of homes with a VCR increases rapidly month by month, as does the range of prerecorded cassettes to play on them.

While the recorder by itself gives you a much greater choice of viewing, used with a camera it provides a new leisure activity – creating your own TV programmes.

1 Video cassette recorders

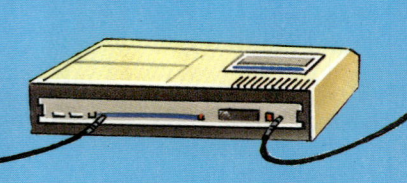

A VCR is the domestic version of the video tape recorder used in TV broadcasting. The tapes are stored in handy cassettes instead of large reels. The lead from the TV aerial plugs into the aerial socket at the back of the VCR and a lead connects it to the TV set. The TV signal passes from the aerial to the VCR.

To record a programme the signal is then transferred to tape by the recording heads of the VCR. The TV does not even have to be switched on. To watch broadcast programmes at the time they are broadcast, the signal comes from the aerial, through the VCR to the TV.

3 VCR features

REWIND: Some machines will automatically rewind the tape when it reaches the end or the end of a programmed recording.

LOADER: This is where you put the cassette in. Some machines load from the front, some from the top. In top loaders you push the cassette into an open drawer and push the drawer down into the machine. Front loaders used motorized loading. The loader grabs the cassette from you and takes it down into the machine.

AUTOMATIC EDIT START: This gives you invisible joins between picture frames when you stop and start the tape. Very useful when you are recording from a camera.

FREEZE-FRAME (STILL FRAME): This allows you to hold a single frame continuously on the screen so that you can view it more closely. On most machines the picture quality is not very good on freeze.

FAST PICTURE SEARCH (FORWARDS AND BACKWARDS): This allows you to run through the tape very quickly (usually about five times the normal speed) while still keeping a picture on the screen. This is very useful for finding the beginning of recordings, missing out commercials and running through boring bits.

SLOW MOTION: Some machines have single speed others have variable speed – you slow the picture down gradually until it reaches the speed you want.

FAST PLAY/DOUBLE SPEED: This usually runs at two or three times the normal speed. It can be used for comic effect or for getting the messages from a programme very quickly.

AUDIO DUB: This allows you to record sound on a previously recorded cassette, so that you replace the existing sound track. Only really useful if you are going to use a camera with your VCR.

FRAME BY FRAME/FRAME ADVANCE: You can advance the picture one frame at a time.

TAPE COUNTER: You set the counter to 0 at the beginning of the tape. You can then use it to help you index your tape and find programmes quickly.

All VCRs allow you to record off the TV while you are watching the programme you want to record, while you are watching another channel, or while the TV is turned off. They have their own tuners built into them, which are completely separate from the tuner which selects the channels on the TV. They also have a timer, which you can set in advance to record at least one programme while you are out.

To do all these things all VCRs have certain basic "features". The more expensive the machine the more special features it has. Here you can see the main features that are available. The symbols used on the machines vary from one model to another.

60

2 VCR formats

There are three main types of VCR – VHS (Video Home System), Betamax and Video 2000 (the Philips system). Within each of these three formats there is a range of models at different prices and offering different facilities.

All three systems use half inch wide tape in their cassettes, but you cannot use cassettes made for one system on machines from either of the other two systems.

There are good machines in all three formats. In deciding which one to choose remember that to swap tapes with friends you must have the same format. It is also a good idea to check which prerecorded programmes are available in the format you have chosen.

Cameras and portable VCRs

With a video camera† you can record pictures and sound on to a cassette in your VCR. If the VCR is connected to a TV you can watch the picture on the screen, while recording it. The camera is linked to the VCR by a long lead. Sometimes you need to use an adaptor between them. If you want to record outside or have more freedom of movement, you need a portable VCR. These will work off batteries as well as mains electricity. You need a separate tuner if you also want to use your portable VCR to record broadcast programmes. A battery charger is very useful and will also act as a mains adaptor.

Tuner

Portable VCR

CHANNEL SELECTOR: You select the TV channel you want by pressing these buttons. Some VCRs have the facility to receive up to 16 TV stations.

TAPE END ALARM: Light warns you when the tape is nearly finished.

CAMERA CONNECTOR: Some machines allow you to plug a lead from a camera straight into the VCR. With other machines you have to use a special adaptor or power supply unit.

CLOCK/TIMER DISPLAY PANEL: The clock shows the actual time. The timer is like an alarm which you set in advance to start the machine when the programme you want to record begins. When you set the timer, the clock disappears and the panel shows the day, time, channel and length of the programme you want to record.

The amount of time in advance you can set your machine to record varies between 12 hours and two or three weeks. "Programmable" or "multi-mode" timers allow you to preset your machine to record several different programmes at different times on different channels in advance (but remember that the tape playing time is usually only 3 or 4 hours).

Video discs

Video discs are similar to long playing records in size and shape. They are much cheaper to make than video tapes, but they cannot be used for home recording, either from the TV or from a camera. All you can do with a video disc player is play prerecorded discs. However, information can be stored and located very accurately on discs. This means that the quality of the picture is usually very good at all speeds. At the moment it seems likely that discs will be used more for business and educational purposes than for home entertainment.

As with VCRs there are different disc systems and each system uses a different type of disc.

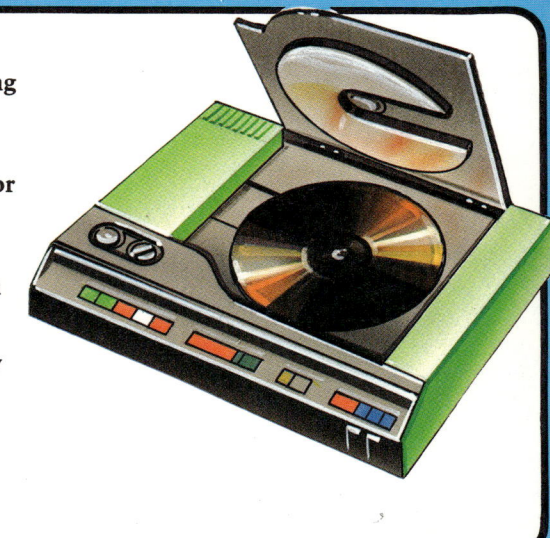

4 Remote controls

Most VCRs now come with remote controls. The cheaper machines have a lead linking them to the remote controls. The more expensive use infra-red rays.* Some just have a pause control, others have all the controls that are on the machine. Most people find that they do not use their trick frame facilities much unless they are available on remote controls.

Tips on using VCRs

1. Make sure that you buy good quality tapes by sticking to well-known makes. If the oxides on the tape are not properly bonded to the backing they can flake off and clog moving parts in the machine.
2. Do not hold the tape on still frame for more than a few minutes. You can clog the heads and damage the tape if you do.
3. Keep cassettes in cases to avoid getting them dusty and dirty.
4. Label your cassettes as soon as you have recorded on to them. It is very easy to lose something you have recorded, especially if you have several tapes.
5. If you have a recording you specially want to keep you can remove the record lockout tab on the tape to make sure you do not accidentally record over it. You can replace the tab later, if you decide to record over it.

*To find out how infra-red remote controls work see page 49   †For more about video cameras see pages 38 and 62

Using a video camera

To take good pictures with a video camera you need to be aware of the same factors as you would think about if you were using a still camera – focusing, adjusting the amount of light entering your camera and careful positioning of your subject within the edges of your picture. But it also helps to be aware of a few things that apply particularly to taking moving pictures. The great advantage with video is that you can experiment as much as you like without wasting any tape, because you can record on it over and over again.

It is important to keep the camera steady while you are shooting – sequences that jump and jerk about are very difficult to watch. It is often a good idea to rest the camera on a firm surface, or, better still, use a tripod, so that you can turn the camera and tip it up and down while shooting. Some cameras are designed to be hand-held, some to rest on the shoulder. When holding and moving with the camera try to move your whole body as smoothly as possible.

Camera shots

Any programme, whether it is made by professionals or amateurs, is composed of a series of separate shots. A shot is one individual scene. Each time you stop the camera running it is the end of a shot. A good cameraman varies his shots to provide visual interest, taking his subject from several different angles, and distances. You get the best results from planning your shots in advance and thinking how one shot will lead into another. Try to make each shot at least 10 seconds long, if they are any shorter the result will be rather jumpy and fragmented. Here are some of the most basic shots to try out and combine with each other.

A long shot gives the general setting without much detail. It is useful for introducing viewers to a subject and for endings.

In a mid shot the main subject and the background have equal importance.

Close-ups have the most impact on the viewer. All the attention is on one specific thing and the background becomes unimportant. Leave enough room for a border area around the subject.

High angle  / **Straight on**  / **Low angle**
Another way of varying your shots is to change the height from which you shoot. The tendency is to shoot from one height (your own or your tripod's) all the time.

Panning is turning the camera from side to side during the shot. Move the camera slowly. A 360 degree pan should take at least one minute. Much faster would confuse the viewer.

◀ With a zoom lens you can move between long shots and close-ups without moving the camera during the shot. Move the zoom slowly and smoothly. Too much zooming can be disturbing to the viewer.

Tilting is tipping the ▶ camera up or down. It is a good idea to hold the camera still for about three seconds when starting to pan or tilt and when completing the shot.

62

TV in the future

Television is likely to continue on its present course of rapid change into the foreseeable future. All sorts of astonishing innovations are already technologically possible. How soon they will come into general use, or whether they do at all, depends on whether anybody wants to buy them and what they are prepared to pay. In the immediate future, cable and satellite broadcasting are likely to bring about a tremendous increase in the number of TV channels. Stereo sound, more compact equipment and changes in the size and nature of the image on the screen are all likely to follow.

The camera tube in TV and video cameras will eventually be replaced by a light-sensitive silicon chip. This would allow cameras to become very much smaller. Eventually all electronic cameras will probably have a tiny built-in video recording system, just as cine cameras have film inside.

The picture tube in TV sets may also become obsolete. This would mean that sets could be much thinner and could be made both larger and smaller than the present range of sets. People are experimenting with various different systems of creating these "flat screen" televisions.

HDTV

The problem with increasing the size of the screen is that as the picture gets bigger, the detail or "definition" decreases. To solve this problem various systems of producing high-definition television (HDTV) are being developed. They all depend upon increasing the number of lines per picture to over a thousand. Very good results can already be achieved, even when the picture is the size of a large cinema screen. However, an HDTV system requires its own special cameras, TV sets and recorders. It would also require new methods of transmission, probably by satellite because it takes up so much room on the air waves.

3D TV

Three-dimensional television (3D TV) makes the images on the screen look as though they are coming out of the TV towards the viewer. There have already been some quite successful experiments in 3D TV. The viewers have to wear special spectacles, which have a red filter for one eye and a green filter for the other. The picture consists of two images – one superimposed on top of the other. One image is red and one image is green. Through the coloured spectacles each eye can only see one image. This fools the viewer into seeing a single three-dimensional scene in black and white.

TV holography

Another way of recreating 3D images is through holography – a technique which uses laser beams and photographic plates. To make TV holography possible scientists will have to find a practical way of capturing and transmitting moving holograms. Research in this area is very complex and expensive and is still in its early stages.

TV communications

Future technology could turn television into one of the main ways for people to communicate with each other. Already video cameras and cable systems have made it possible for groups of people in different places to hold "meetings" in which they can all take part. Perhaps one day, instead of telephoning someone, you will simply summon their image to the screen and talk into the TV set.

TV and video words

Here is a list of some words and phrases, which are used in conection with TV and video.

BANDWIDTH A range of radio frequencies or wavelengths, falling between two limits.

CATHODE RAY TUBE Tube capable of producing electrons and focusing them in a beam. Both camera tubes and picture tubes inside TV sets are cathode ray tubes.

CHANNEL A band of radio frequencies wide enough for the transmission of radio and television signals.

CLOSED CIRCUIT TV (CCTV) Any form of television that is not broadcast to the general public. Closed circuit systems are often used in schools, offices, security systems, traffic surveillance, etc. The signals usually travel along cables.

DIRECT BROADCAST SATELLITE (DBS) A satellite capable of transmitting radio waves strong enough to be received by a small dish on the roof or in the garden of a private house.

DUB Make another sound track, or add sound (often music or sound effects) to an existing sound track. It can also mean to copy a recorded sequence or programme.

ELECTRONIC NEWS GATHERING (ENG) The use of video cameras and portable video recorders to collect news stories for TV programmes.

ELECTRONIC FIELD PRODUCTION (EFT) Recording a programme on location, generally using one video camera and one portable recorder.

FORMATS The different types of recording system used in home video equipment.

FREQUENCY The number of radio waves per second.

GENERATION The number of recordings away from an original recording. A first generation tape is a recording made from an original tape. A second generation tape is a recording made from a first generation tape, and so on. The quality of the picture and sound gets worse with each generation.

HARDWARE Equipment used in making or displaying TV and video programmes, as opposed to "software", which describes the programmes themselves. The terms "hardware" and "software" originated in the computer industry.

MASTER An original tape or disc from which copies are taken.

NETWORK A group of broadcasting stations connected for the simultaneous broadcast of the same programme.

ON AIR A programme is "on air" when it is being broadcast.

ON LOCATION Away from a studio. A programme made "on location" is filmed or video taped in some situation other than a TV or film studio.

OUTSIDE BROADCAST (OB) A programme made by taking TV cameras and control room equipment to some location other than a studio.

PORTAPACK A portable video system consisting of a video camera and a battery-powered video recorder.

POST-PRODUCTION The final stage of a production, in which the editing and dubbing take place.

RECORDING OFF-THE-AIR Recording a broadcast programme on video tape.

SCANNING The process of the electron beam moving across the target plate of the camera tube, or the back of the screen in the picture tube of a TV set.

SHOT The term used to describe each individual scene taken by a TV or film camera.

SOFTWARE Programmes on video cassettes or discs, as opposed to "hardware", which describes the equipment used in making or playing back TV and video programmes.

TELETEXT A system of transmitting information from a central computer information bank to a TV set by broadcasting the information alongside the normal TV signals.

TRICK-FRAME The term used for the range of special effects, such as freeze-frame, offered by many VCRs.

UHF (ULTRA HIGH FREQUENCY) A range of radio frequencies used to carry TV signals.

U-MATIC (OR U-FORMAT) The semi-professional ¾ inch video tape format, usually used in industry and education. The tape is wrapped round the recording head drum in a U-shaped pattern.

VHF (VERY HIGH FREQUENCY) A range of radio wave frequencies used to carry TV signals.

VIEWDATA (VIDEOTEX) A system of transmitting information from a central computer information bank to a TV set by sending signals along telephone wires.

WAVELENGTH The length of one complete radio wave.

AUDIO & RADIO

John Hawkins and Susan Meredith

Contents

66	The world of audio	84	How audio equipment works
68	How radio works	86	The controls on audio equipment
70	Making a radio programme	88	Make your own radio
72	What are radio waves?	91	CB and two-way radio
74	Transmitting sound	92	Different uses of radio and audio
76	Inside a radio	94	The future
78	How cassettes and records are made	94	Audio and radio words
82	Audio equipment	95	Index

Audio and Radio was written by John Hawkins and Susan Meredith and illustrated by Jeremy Banks, Graham Round, Graham Smith, Rex Archer, Joe McEwan, Mike Roffe, Philip Schramm and Martin Salisbury.
Designers: Round Designs and Roger Priddy

The world of audio

Audio is really just another word for sound, and audio equipment is anything which reproduces sound so that we hear it through loudspeakers or headphones.

First of all, the sound has to be converted into electricity so that it can be stored and then reproduced in a different place from where it was made.

A radio is a specially complicated piece of audio equipment. It reproduces sound which has travelled as electricity, not only down wires but also through the air, on invisible waves. The sound which comes out of a radio can be either live (being made as you hear it) or pre-recorded.

Audio equipment is often called hi-fi. Hi-fi is short for "high fidelity", which means the equipment is designed to reproduce the original sound as exactly as possible.

1 What is sound?

When someone makes a noise, tiny particles of air nearby vibrate backwards and forwards. These particles make your eardrums vibrate and produce tiny pulses of electricity, which travel along nerves to your brain. Your brain recognizes the pulses as sound. The vibrating air particles are called sound waves. It is important to remember that these are different from the waves used for radio.

2 Hands in line with speaker.

You can sometimes feel sound waves as well as hear them. A loud noise, such as thunder, can make the air vibrate so much that things shake. To feel the sound coming from a loudspeaker, try blowing up a balloon until it is firm, then hold it lightly between your hands within a metre of the speaker. Have your hands in line with the speaker. You may need to move the balloon about a bit to feel the vibrations.

1 Radio

Telephone exchange

Before radio was invented, the only way of making sound travel over long distances was by telephone. The phones of two people having a conversation need to be linked by wires for electricity to flow along.

Transmitting aerial

Transmitter in here

2

For radio, there has to be a transmitter for sending out the sound and a receiver for hearing it. These work by electricity but they are not joined to each other in any way. The sound travels from the transmitter on invisible waves which "radiate" from a transmitting aerial.

Receiver

How sound becomes electricity

The first step in recording and reproducing sound is to convert it into electricity. This is done by a microphone. There are different sorts of microphone. Moving coil types, like this one, work by magnetism.

1 Lattice-work lets sound waves through.

2 When sound waves hit this flat disc, or diaphragm, they make it vibrate. The diaphragm vibrates at the same speed, or frequency, as the sound waves.

3 This coil of wire is attached to the diaphragm. When the diaphragm vibrates, so does the coil.

4 This is a magnet. Whenever a coil of wire moves near a magnet, an electric current is produced. This is what happens in the microphone.

5 The electric current is called a sound signal. It varies according to the frequency and loudness of the sounds. It goes down a wire connected to the coil.

3

A high note, say from a recorder, makes the air vibrate faster than a low note, say from a tuba. So, high-pitched noises are said to have a higher "frequency" than low ones. We can only hear sounds within a certain frequency range. Animals often hear things we cannot.

How radio works

These two pages will give you a general idea of the way radio works. They explain how a radio programme gets from the studio of a large radio station to your radio at home. Later in the book you can find out more about each stage in the process and about the different ways radio is used.

The studio

The people taking part in the programme talk into microphones in a studio. The sounds of their voices are converted by the microphones into electric signals. These sound signals travel through wires to the studio control room, where they are strengthened, or amplified, in the control desk. Studio engineers listen to the sounds from the studio through headphones or loudspeakers. They also have meters on the desk, which show them the sound levels. They control the volume and tone of the sounds by moving knobs and switches.

Studio
Control desk
Tape recorder
Studio control room

Programmes from other studios.
Signals from studio to continuity desk.

The continuity desk

The signals then go through wires to the continuity desk. The people here have the job of linking programmes together in the right order. They select signals first from one studio, then from another. These include signals coming through wires from tape recorders, when programmes are not being broadcast live but have been pre-recorded in the studio. Continuity announcers do the talking between programmes and insert things like time-signals and jingles.

The control room

From the continuity desk, the signals go to the control room. Large radio stations broadcast several programmes at the same time on different channels. Each channel has its own continuity desk and the control room has to sort out the signals from each one. From the control room, engineers send the signals out to transmitters, usually through cables similar to those used for telephones.

Signals from continuity desk to control room.
Control room
Signals from other continuity desks.

The transmitter

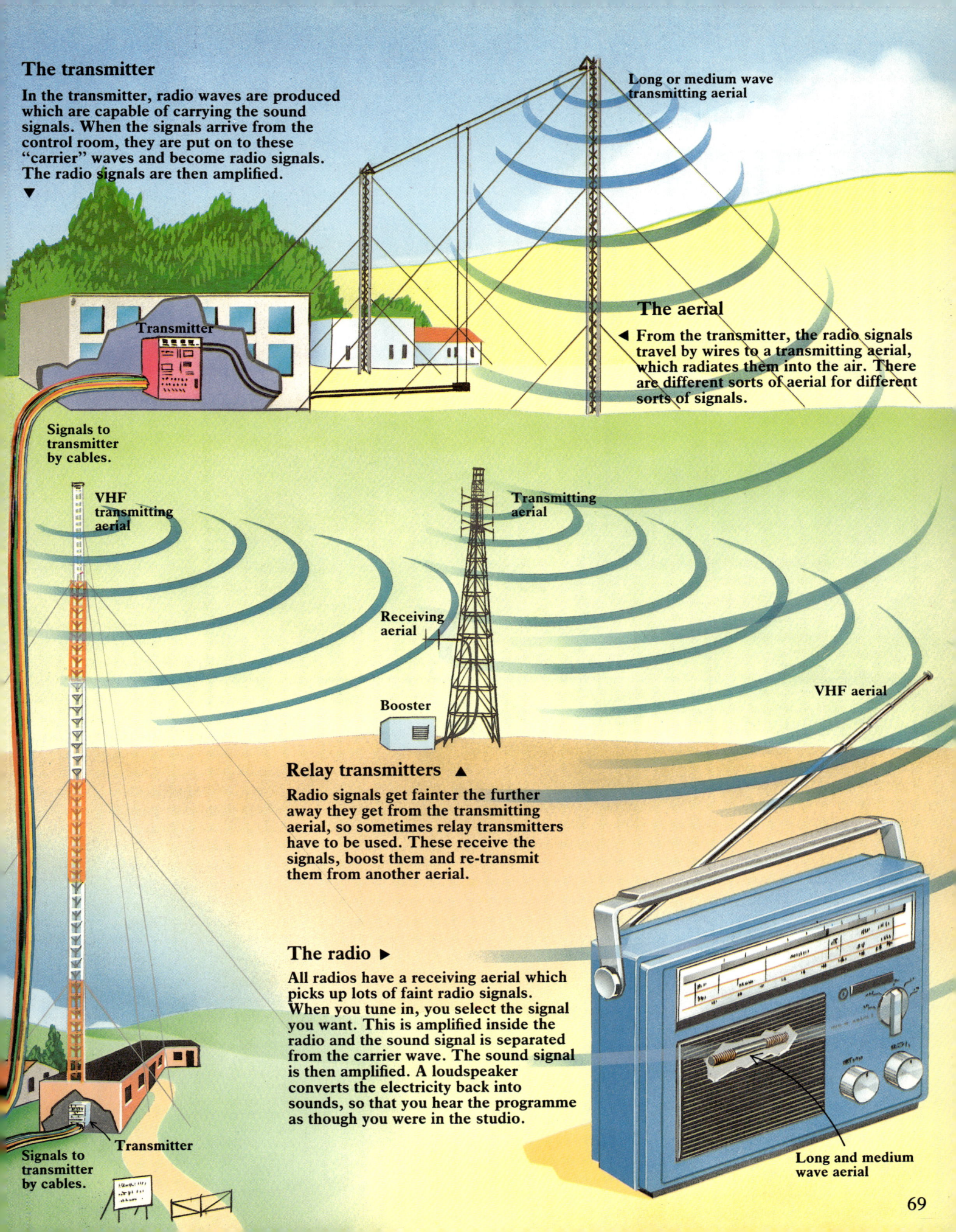

In the transmitter, radio waves are produced which are capable of carrying the sound signals. When the signals arrive from the control room, they are put on to these "carrier" waves and become radio signals. The radio signals are then amplified. ▼

Long or medium wave transmitting aerial

Transmitter

The aerial

◀ From the transmitter, the radio signals travel by wires to a transmitting aerial, which radiates them into the air. There are different sorts of aerial for different sorts of signals.

Signals to transmitter by cables.

VHF transmitting aerial

Transmitting aerial

Receiving aerial

Booster

VHF aerial

Relay transmitters ▲

Radio signals get fainter the further away they get from the transmitting aerial, so sometimes relay transmitters have to be used. These receive the signals, boost them and re-transmit them from another aerial.

The radio ▶

All radios have a receiving aerial which picks up lots of faint radio signals. When you tune in, you select the signal you want. This is amplified inside the radio and the sound signal is separated from the carrier wave. The sound signal is then amplified. A loudspeaker converts the electricity back into sounds, so that you hear the programme as though you were in the studio.

Signals to transmitter by cables.

Transmitter

Long and medium wave aerial

Making a radio programme

Some radio programmes are fairly straightforward to organize. For discussions, chat shows, panel games and even plays, the people taking part can sit round a table in a studio, talking into microphones. A news and record programme, like the one shown in these pictures, is more complicated. So many different bits have to be incorporated, including pre-recorded snippets and items produced outside the studio, that split-second timing and co-ordination are essential.

Very accurate clocks help the DJ time the programme exactly.

Soundproof window between studio and control room.

Control room

If the music is on tape, it is sent out by the engineer at the right time.

This is a "phone selector", used for phone-in programmes.

Faders

Faders control volume.

DJs can talk to the listeners or the engineer through their microphones, depending on how they set their microphone switch. For "voice-overs", the music is automatically faded down whenever the DJ speaks.

Preparing a news and record programme

The producer, disc jockey, engineer and production secretary meet about a week before the programme to plan any special items, such as interviews with famous people. The day before the programme, they finalize details such as which records to play, which dedications to read out and the exact timing of the programme.

The news

Sometimes the DJ reads the news, sometimes it goes from a separate studio direct to the engineer's desk. The DJ plays a jingle to announce the news and the engineer "fades up" a channel on the desk, linked to the newsroom.

On-the-spot reporters

On-the-spot reporters do interviews using high quality, portable recorders. The tapes are taken back to the studio and sent out from the control room. They can be "edited" to make them the right length.

After going through the engineer's desk, the signals leave the control room through cables in this "jackfield". Items from outside the studio arrive at the desks through the jackfield.

The engineer feeds the sound back from the control desk to the DJ's headphones and can talk to the DJ through a microphone.

Engineer's control desk

DJ's control desk

Delay box for phone-in programmes.

Pre-recorded cassettes of jingles, adverts or sound effects are played in this "jingle box".

There are two turntables. While one record is playing, the DJ gets the next one ready on the other turntable, so it can be switched on as soon as the first record ends.

In the DJ's studio

Disc jockeys are responsible for linking the different parts of their programme smoothly together and for pacing the programme correctly. They have a rough script, agreed with the producer beforehand, so they know what order to do everything in, but they also improvise a lot. The signals go from the DJ's control desk to the engineer's desk. Engineers make sure the signals go to the transmitter without any technical hitches. They control the volume and tone of the sounds. Items from outside the studio also go through the engineer's desk before going to the transmitter.

Phone-ins

When listeners phone in, the producer decides whose calls to include. The production secretary phones the callers back when they are wanted on the air. The calls reach the DJ through the "phone selector", which is like a small switchboard. The microphone and headphones act as the DJ's telephone. The callers' words are delayed by about a second so the engineer can cut out anything abusive.

Remote studios

Speakers convert signals from studio into sound.

Signals go from one studio to other by cables.

Engineer

Remote studios are used for interviews at places like airports. The signals have to go through the control desks in the usual way, so the remote studio is linked to the main studio by wires. The link is two-way, so the people being interviewed can hear the DJ. The control room engineer and the remote studio engineer can communicate with each other by telephone.

Radio cars

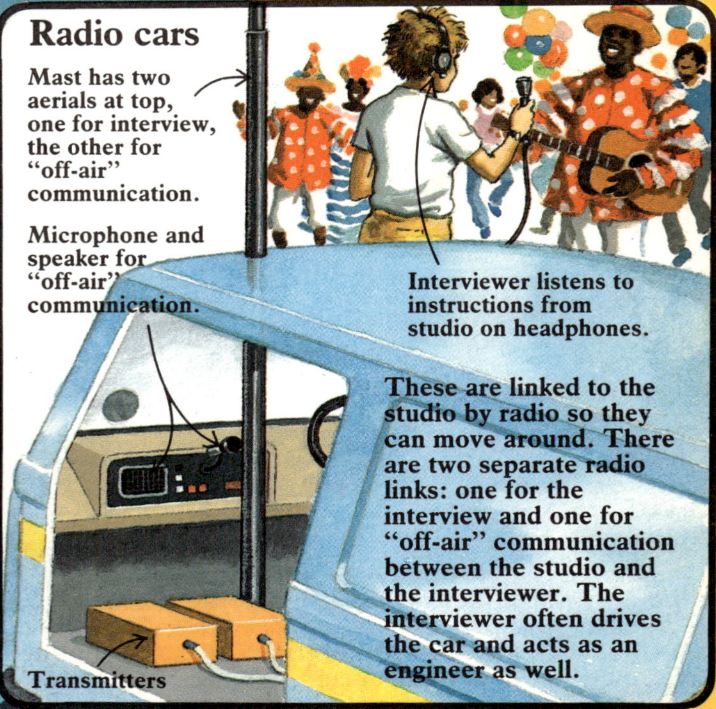

Mast has two aerials at top, one for interview, the other for "off-air" communication.

Microphone and speaker for "off-air" communication.

Interviewer listens to instructions from studio on headphones.

These are linked to the studio by radio so they can move around. There are two separate radio links: one for the interview and one for "off-air" communication between the studio and the interviewer. The interviewer often drives the car and acts as an engineer as well.

Transmitters

71

What are radio waves?

Radio waves are pulses of electrical energy, which can travel through air, space and even solid objects. You will sometimes hear them referred to as electromagnetic waves. Some radio waves come from stars or are created by lightning but these cannot be used to carry sound. To carry sound, the waves have to be made in a transmitter.

The term "radio waves" sometimes means the waves made in the transmitter, before they have any sound on them. (These are also called carrier waves.) Or it can mean the waves which go out from the aerial carrying sound.

1 When you throw a stone into a pond, circles of ripples spread outwards from it. You can imagine radio waves looking and moving like these ripples.

2 Radio waves usually travel from transmitting aerials in circles, like the ripples on the pond, but they can also be made to travel in beams.

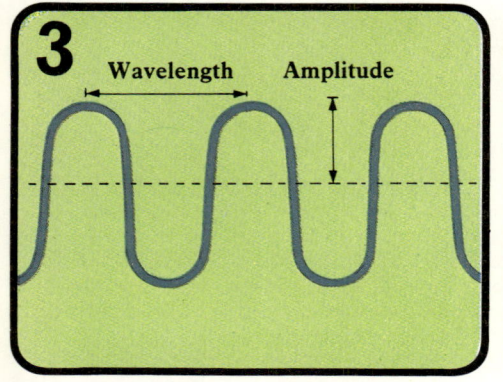

3 The wavelength of waves used for radio varies between 0.33mm and 30km. The distance is measured from the crest of one wave to the crest of the next. The height of a wave is called its amplitude and indicates its strength.

4 All radio waves travel at the speed of light (300,000km per second). This means that the shorter the wavelength the greater the rate, or frequency, at which complete waves go past a certain point. Frequency is measured as the number of times every second a wave goes by. (Because they go so fast, the diagram above shows waves going by in a millionth of a second.) Different frequencies are used for different sorts of radio transmission.

1 Make your own radio waves

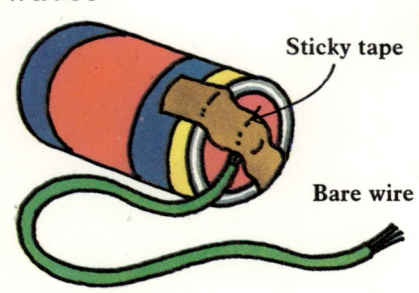

You can make and transmit radio waves to your radio. Using a sharp knife, carefully strip about 2cm of the plastic off both ends of a piece of electrical wire about 40cm long. Fasten one end firmly to either end of a cylindrical shaped battery with sticky tape.

2

Sharpen a pencil at both ends. Twist the loose end of wire neatly round one of the points and fasten it firmly with sticky tape. Tune a portable radio to part of the long or medium waveband where there is no programme and put the battery near it.

3

Moving the pencil and battery round the radio, tap the free pencil point against the free end of the battery. You will hear clicks on the radio as it picks up the radio waves you are making. If you move away from the radio, the signals will get fainter.

How radio waves were discovered

The first person to prove that he had produced radio waves was the German scientist, Heinrich Hertz, in 1887. He set up two metal spheres with a small gap between them and, at the other side of his laboratory, a wire loop with a gap in it. When he used an electric current from a battery to make a spark jump the gap between the metal spheres, he noticed that a spark appeared immediately in the gap in the wire loop. He realized that he had created and transmitted a radio wave from the spheres to the loop.

Radio waves similar to the ones produced by Hertz are made whenever the electric current in a wire changes suddenly, for instance, when electrical equipment is switched on or off. Sometimes the waves show up as interference on a nearby radio or TV.

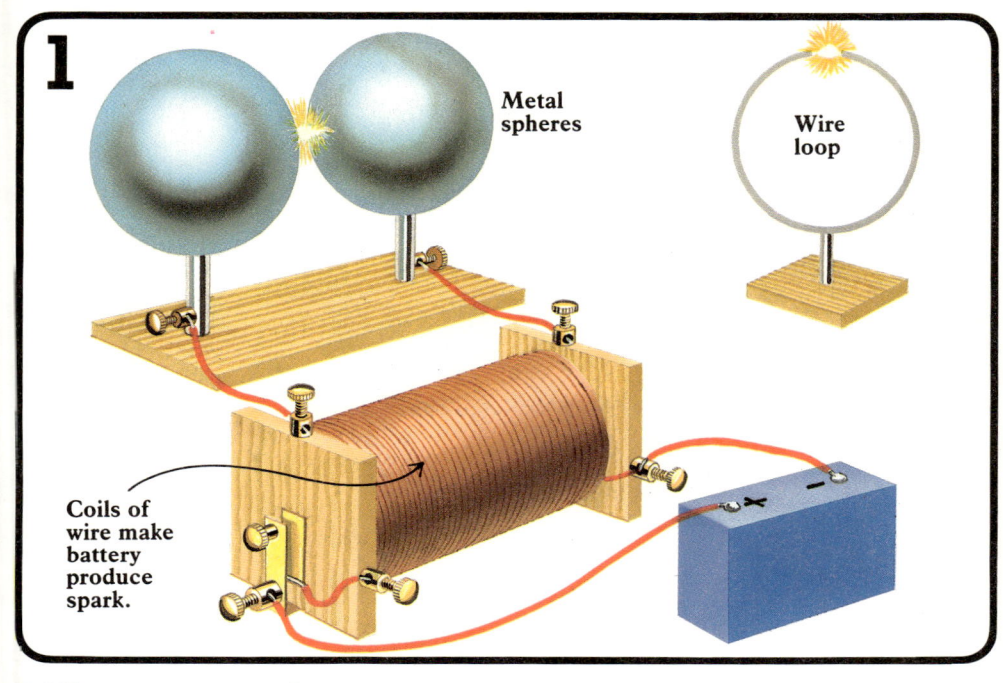

1 Metal spheres — Wire loop — Coils of wire make battery produce spark.

2

Different types of waves

Radio waves are part of a family of waves known as the electromagnetic spectrum. This includes various light and heat waves. The only difference between the waves is their length. Radio waves are the longest. None of our senses can detect them. We feel shorter, infra-red waves as heat. At shorter lengths still, our eyes detect the waves as light of different colours. And at even shorter lengths, there is ultra-violet light, which is used in sun-ray lamps. X-rays and gamma rays are also part of the electromagnetic spectrum. Sound waves are not.

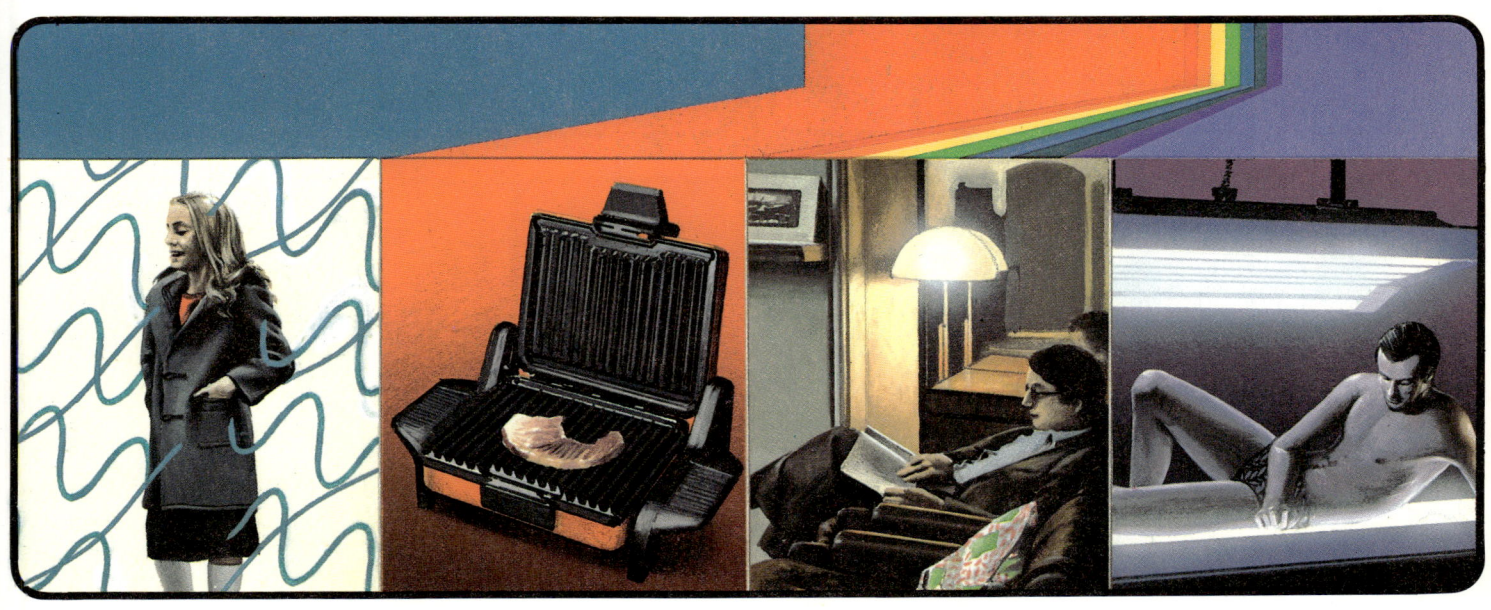

Radio waves go through our bodies without our knowing.

Infra-red rays can be used for cooking.

Visible light is reflected into our eyes to make us see.

Ultra-violet light can be used to produce a sun-tan.

73

Transmitting sound

To carry speech or music, radio waves have to be continuous, not like the ones made by early radio scientists such as Hertz, which only occurred in short bursts. Continuous carrier waves are made, or generated, in radio transmitters. The waves then have the sound signals put on to them in a process called modulation. You can find out about the two different sorts of modulation below. After modulation, the waves are fed through wires to an aerial and from the aerial they are radiated into the air for your radio to pick up.

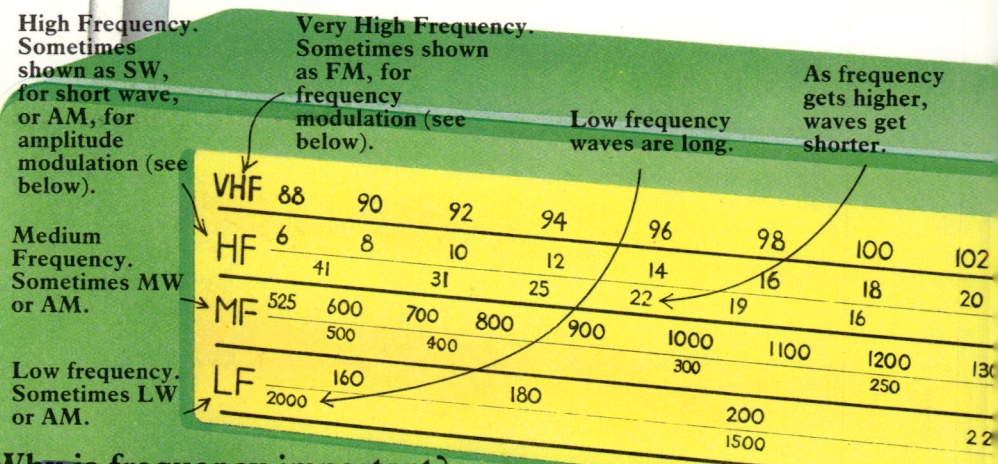

High Frequency. Sometimes shown as SW, for short wave, or AM, for amplitude modulation (see below).

Very High Frequency. Sometimes shown as FM, for frequency modulation (see below).

Low frequency waves are long.

As frequency gets higher, waves get shorter.

Medium Frequency. Sometimes MW or AM.

Low frequency. Sometimes LW or AM.

Why is frequency important?

Radio waves have to have different lengths and frequencies, or they would get mixed up together and it would not be possible to hear one broadcast separately from another. Certain frequencies are reserved by international agreement for use by different radio stations. Other frequencies are kept for other users.

You can see from the radio tuning display above how the waves are divided into bands. This radio has low, medium, high and very high frequency bands. On some displays, low, medium and high frequency waves are referred to by their length (long, medium and short wave). Wavelength is measured in metres. Frequency is measured in kilohertz

1 Putting sound on the waves

Continuous carrier wave.

Signal from microphone.

Amplitude modulated wave.

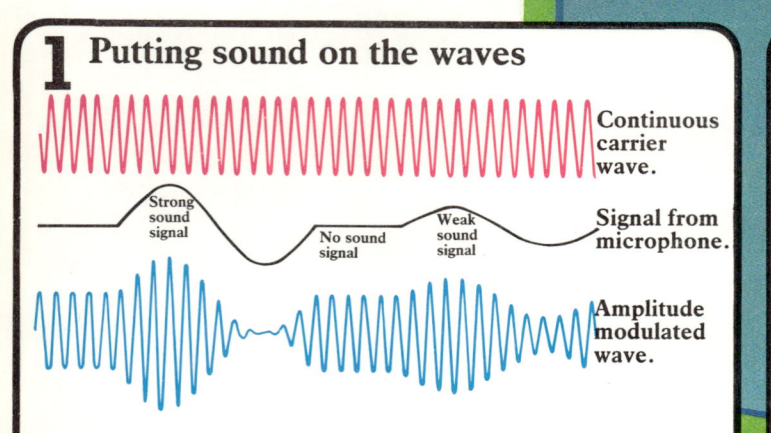

One way of putting sound on radio waves (modulating them) is to alter their amplitude (height). When the signal from the microphone is strong, the amplitude varies a lot. When the signal is weak, the amplitude varies less. This is called amplitude modulation (AM). Low, medium and high frequency waves are amplitude modulated.

2

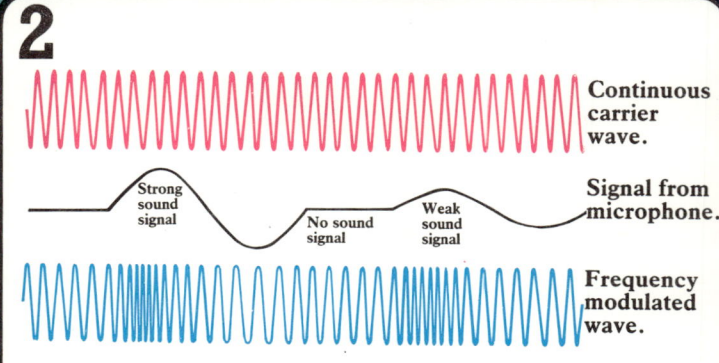

Continuous carrier wave.

Signal from microphone.

Frequency modulated wave.

The other way of putting sound on the waves is called frequency modulation (FM). When the signal from the microphone is strong, the frequency varies a lot. When the signal from the microphone is weak, the frequency varies less. The amplitude of the waves stays the same all the time. VHF waves are frequency modulated.

1 How the waves travel

2

3

Long and medium waves keep fairly close to the Earth, following its shape. Long waves can travel for over a thousand kilometres, medium waves for several hundred, before they lose their energy and eventually fade away.

Short waves go up to a layer in the Earth's upper atmosphere called the ionosphere. This reflects the waves back to Earth, so they travel a very long way. For this reason, short waves are used for overseas broadcasting.

Very high and ultra-high frequency waves do not go through solid objects very well, so they are used for broadcasting over short distances (up to about 150km). They also go off into space right through the ionosphere.

1MHz (megahertz) = 1,000kHz.

106	108 MHz
24	26 MHz
	11 metres
1500	1600 kHz
200	metres
240	kHz
	1200 metres

Sometimes shown in kHz with last one or two figures missing.

Sometimes shown with last one or two figures missing.

Frequency

Wavelength

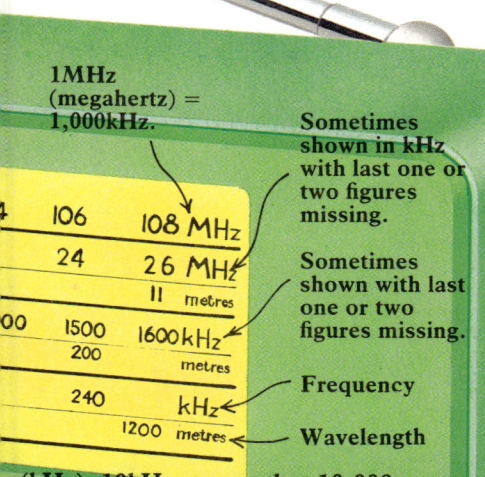

(kHz). 10kHz means that 10,000 waves go past a certain point in a second. The frequency of waves used for broadcasting varies from 10kHz to about 12 million kHz, or 12,000 megahertz (MHz). Sometimes, if there is not much space on the display, the last figure or two of the kHz is missed off.

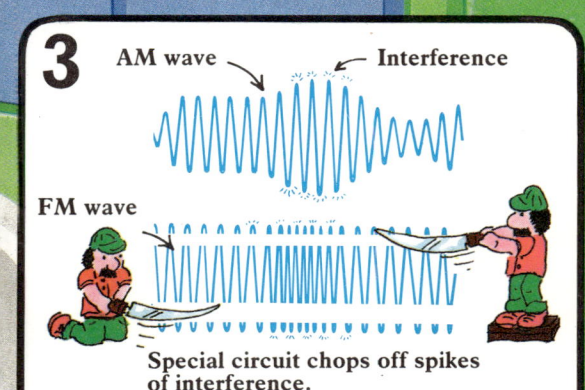

3
AM wave — Interference
FM wave
Special circuit chops off spikes of interference.

Amplitude modulated waves often get "spikes" of interference on them. These are caused by people using electrical equipment and are heard as clicks and crackles on the radio. Frequency modulation cuts out interference but long and medium waves cannot usually be frequency modulated.

How a transmitter works

A radio transmitter has three main parts. An "oscillator" generates the carrier waves. A "modulator" puts the sound signals on them. An "amplifier" boosts the modulated waves. All three parts consist of lots of electronic components arranged in circuits. Here is a simplified picture of an AM transmitter.

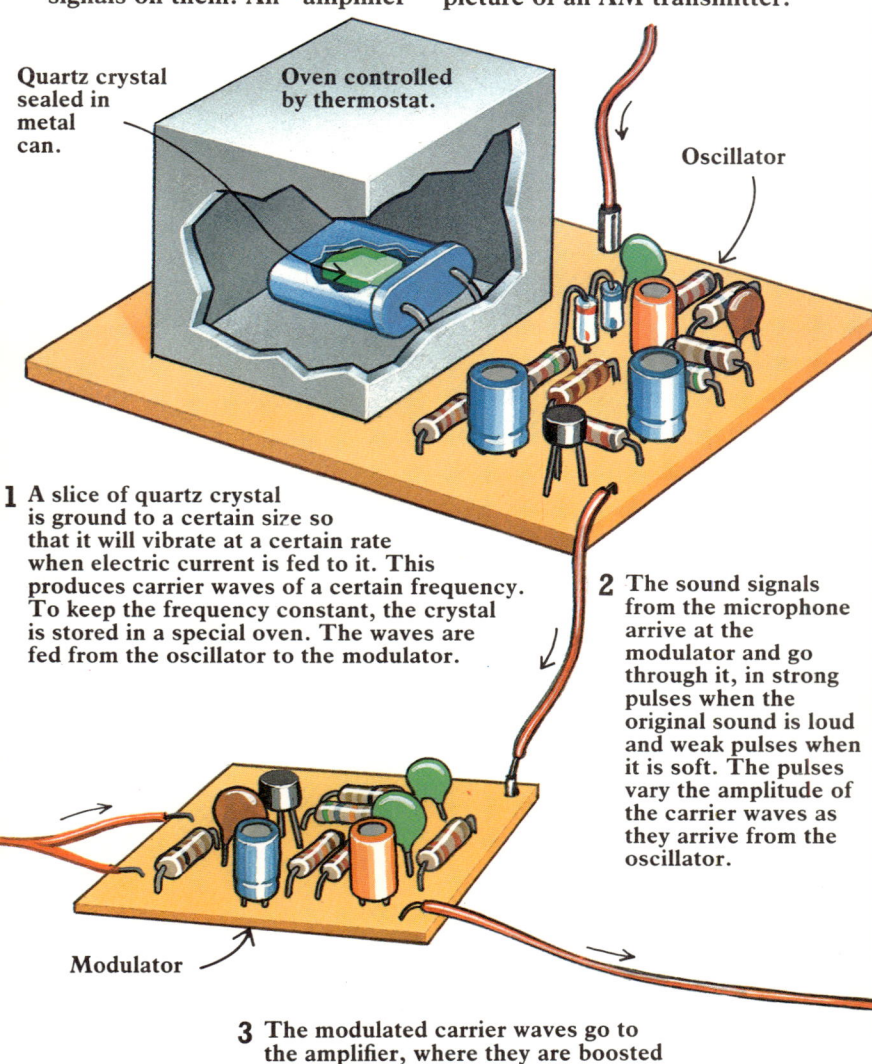

Quartz crystal sealed in metal can.
Oven controlled by thermostat.
Oscillator

1 A slice of quartz crystal is ground to a certain size so that it will vibrate at a certain rate when electric current is fed to it. This produces carrier waves of a certain frequency. To keep the frequency constant, the crystal is stored in a special oven. The waves are fed from the oscillator to the modulator.

2 The sound signals from the microphone arrive at the modulator and go through it, in strong pulses when the original sound is loud and weak pulses when it is soft. The pulses vary the amplitude of the carrier waves as they arrive from the oscillator.

Modulator

3 The modulated carrier waves go to the amplifier, where they are boosted before being sent to the aerial.

4

Super-high frequency waves go through the ionosphere into outer space and are used for satellite broadcasting. The signals can be beamed up from one country and then sent down by the satellite to several others at once.

Getting good reception

Long and medium waves travel best over water or damp ground. If you live in an area with clay soil or lots of rivers, you will probably get good reception of programmes on these wavebands. Reception is usually worst in areas where the rocks are very old.

Picking up more stations at night

During daylight, some of the energy from medium waves goes into space but after dark it gets reflected back from the ionosphere. This means you often pick up medium wave foreign programmes at night that you cannot hear in the day, and you also get more interference.

75

Inside a radio

Your radio is technically called a radio "receiver", because it receives the signals sent from the transmitter. Most of the work in a radio is done by lots of electronic components like the ones in the picture below. There are usually more of each type of component than are shown here. They are arranged in circuits. On the opposite page you can find out what happens inside your radio.

Telescopic aerial for VHF.

Tuning control moves cord over pulleys to operate tuner and move pointer along scale.

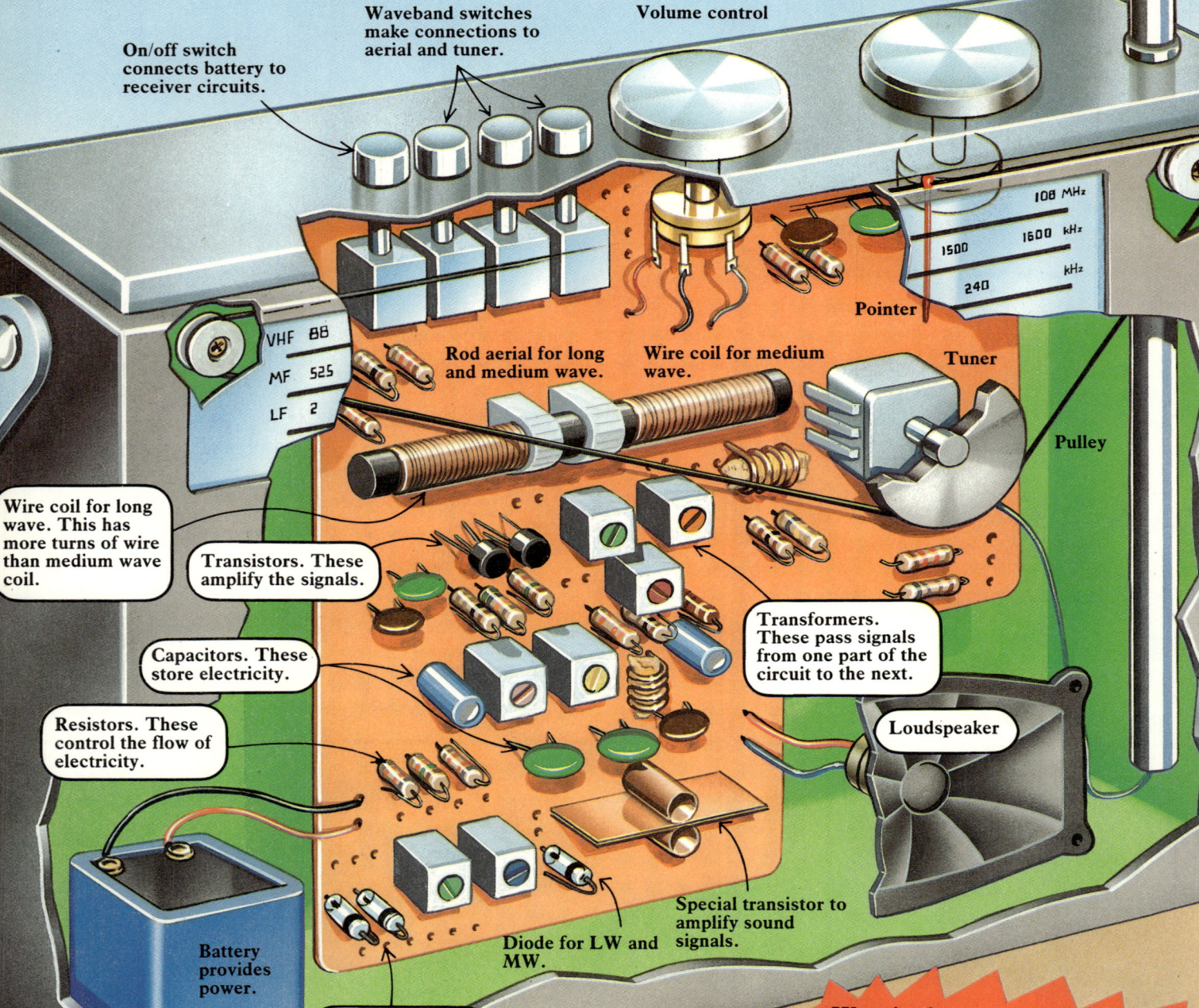

Waveband switches make connections to aerial and tuner.

Volume control

On/off switch connects battery to receiver circuits.

108 MHz
1500   1600 kHz
240   kHz

Pointer

VHF 88
MF 525
LF 2

Rod aerial for long and medium wave.

Wire coil for medium wave.

Tuner

Pulley

Wire coil for long wave. This has more turns of wire than medium wave coil.

Transistors. These amplify the signals.

Transformers. These pass signals from one part of the circuit to the next.

Capacitors. These store electricity.

Resistors. These control the flow of electricity.

Loudspeaker

Special transistor to amplify sound signals.

Diode for LW and MW.

Battery provides power.

Diodes for VHF. These separate the sound signals from the carrier waves.

Warning!
It is dangerous to open up a mains radio. Even if it has been switched off and unplugged for some time, it may still have dangerous charges of electricity stored inside it. If you want to look inside battery-powered equipment, take the batteries out first.

76

How your radio works

Your radio has to pick up the waves coming from the transmitting aerial and separate the sound signals from the carrier waves so they can be converted back into sound that you can hear. When you look inside a radio, it is hard to imagine the different stages in this process. The balloons and robots in this picture should help you understand what happens to the signals on their way from the radio aerial to the loudspeaker. The balloons are carrier waves and the notes are sound signals.

Aerial picks up mixture of waves.

Mixer changes frequency of wanted waves.

Detector diode

Sound signal

Waves amplified

Waves of new frequency only allowed through.

Sound signals amplified

1 When you switch the radio on, the battery or mains electricity provides power to make the receiver circuits work. If you tune to the medium waveband, medium waves passing the set produce a mixture of tiny electric currents in the medium wave aerial coil. Each current corresponds to one from a transmitter. If you switch to the long waveband, the same thing happens in the long wave aerial coil.

2 If the radio is switched to the VHF band, VHF waves produce tiny currents in the telescopic aerial rod.

Tuner throws away most of unwanted frequencies.

3 When you turn the tuning control to a particular spot on the waveband, waves of the frequency you want are allowed to pass from the aerial through the tuner and into the "mixer". Waves of other frequencies are almost all rejected.

4 There are still a few unwanted frequencies left. These have to be got rid of or you will hear bits of other broadcasts as well as the one you want. The carrier frequency you want is altered to a "fixed" or "intermediate" frequency, so that only it can go on to the next stage in the process, through the "fixed tuned circuit". It is altered by mixing it with waves made in a small "local oscillator".

5 The fixed tuned circuit only lets through the waves you want. These are then amplified.

6 The amplified signals go to a "detector diode", which separates the sound signals from the carrier waves (demodulates the carrier). The carrier waves are then thrown away.

7 The sound signals are amplified to make them strong enough to drive the loudspeaker. The amount by which the signals are amplified depends on where you set the volume control.

8 The loudspeaker converts the signals back into sound. You can find out how loudspeakers work on page 85.

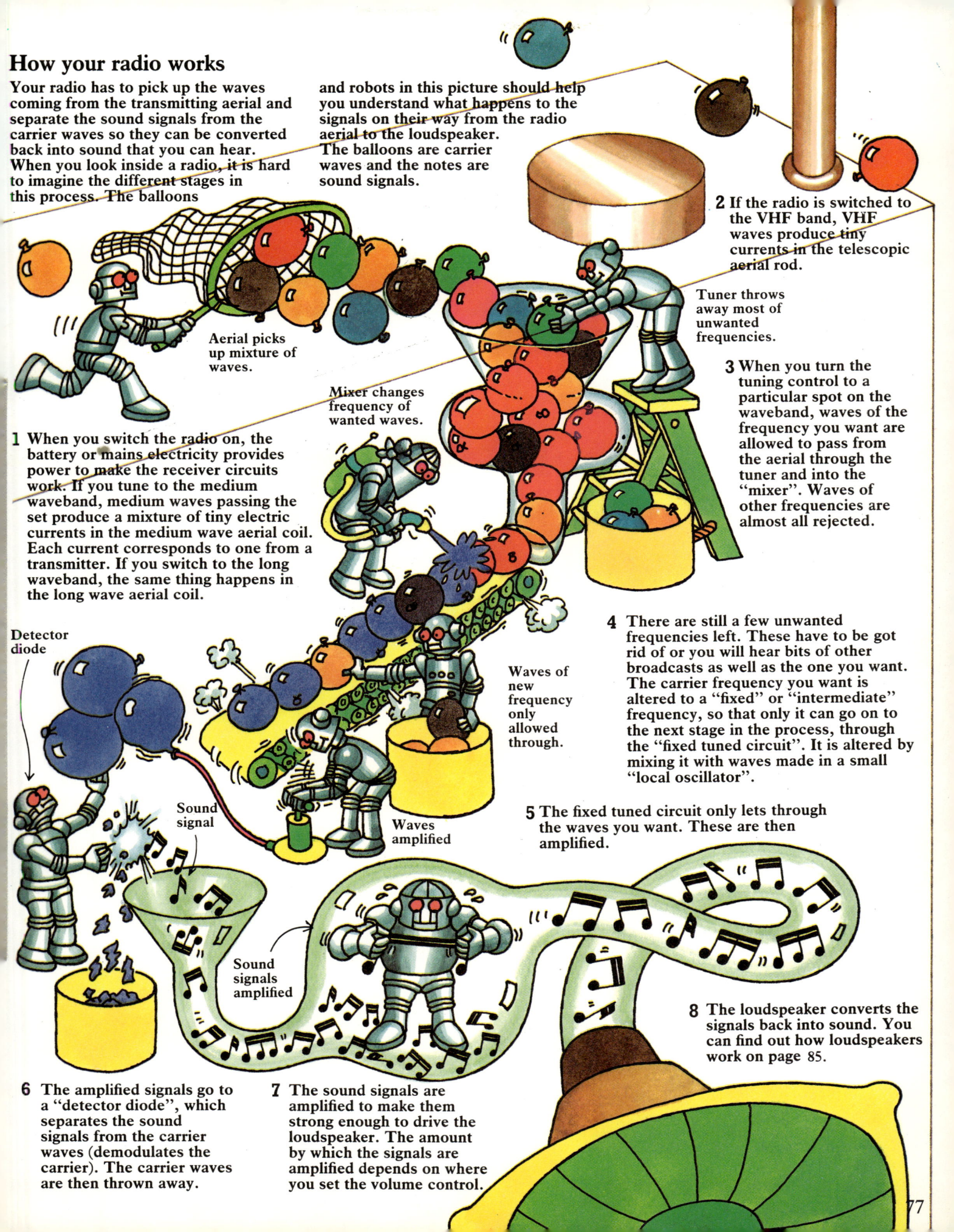

How cassettes and records are made

On the next four pages you can find out how pop cassettes and records are made. Both cassettes and records are usually made from a series of tapes, which are recorded in a studio. Tape is used because it can be corrected, or edited, fairly easily.

The studio

The musicians play into microphones in a soundproofed studio. They are separated from each other by screens, so the sounds from one instrument are not picked up by anyone else's microphone. The microphones convert the sounds into electric signals and these go down wires to a control desk, or mixer, in the control room. The musicians wear headphones so they can listen to the overall sound, which is fed back to them from the control room.

How sound gets on to tape

1 The part of a tape recorder which puts the sound on tape is called the recording head. The unrecorded tape is fed to the head from this feed spool.

2 The tape is made of plastic. It is coated with millions of tiny, invisible particles of a substance called iron oxide.

3 The sound signals come from the microphones, through the desk, to these coils of wire in the recording head. Each coil of wire is wrapped round an iron block. There has to be a separate coil and block for each track on the tape (see opposite page). The current in the wires varies according to the sounds.

4 The current in the wires turns the iron blocks into magnets. Their magnetic strength varies with the strength of the current.

The desk

The desk has lots of identical channels, or modules, one for each microphone. The engineer can control the signals from each microphone separately. Here is a typical module.

The VU meters measure volume units and show the engineer how loud the sound is.

This knob cuts out an instrument if it is not needed for part of the piece. Or it can "solo" the instrument for the engineer to check, by lowering the sound levels of all the other instruments.

These frequency knobs alter the tone of the sounds. For example, if a bass guitar sounds too booming, its low notes can be made softer and its high notes louder.

Echo knob (see page 80).

Stereo pan pot (see page 80).

This "fader" slides up and down to alter the volume of the sounds.

The control room

The producer and sound engineer listen to the sound in the control room, through loudspeakers or headphones. The engineer amplifies the sound signals by moving knobs and switches on the desk. The producer can give instructions to the musicians by talking into a microphone. The musicians hear the instructions on their headphones.

6 The tape now has sound stored on it in a magnetic pattern. It goes to the take-up spool. (You can find out how the sound is played back on page 85.)

5 As the tape passes across a tiny gap at the front of each coil and block, the particles of iron oxide on the tape are magnetized by the blocks. Their magnetic strength varies with the strength of the current and the magnetic strength of the blocks.

Correcting tape

The tape can be fed past an "erase head" in the recorder. A strong electric current demagnetizes the tape, removing the sound so a new recording can be made. Tape of certain widths is also edited by having sections cut out and the ends rejoined.

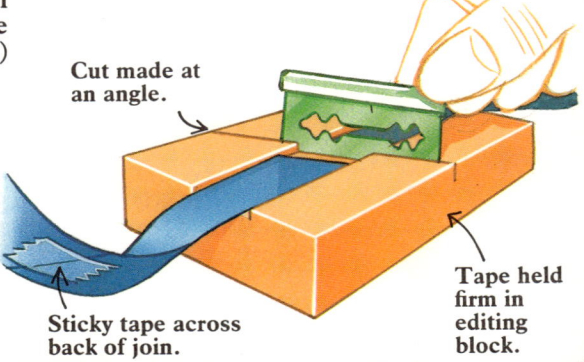

Cut made at an angle.

Sticky tape across back of join.

Tape held firm in editing block.

The recorder

The signals do not go to the tape recorder until the producer is satisfied with the sound. Then they are sent down wires from the desk. The first recording is usually made on tape which is two inches wide. Up to 24 tracks are recorded running parallel to each other across the tape. Sometimes each microphone has its own track. Sometimes the signals from two microphones are combined. The tracks do not have to be recorded at the same time, so the musicians do not all have to be at the recording session together. If someone makes a mistake, their track can be re-recorded separately. This is called overdubbing.

79

HOW CASSETTES AND RECORDS ARE MADE

The master tape

Echo added through this channel.

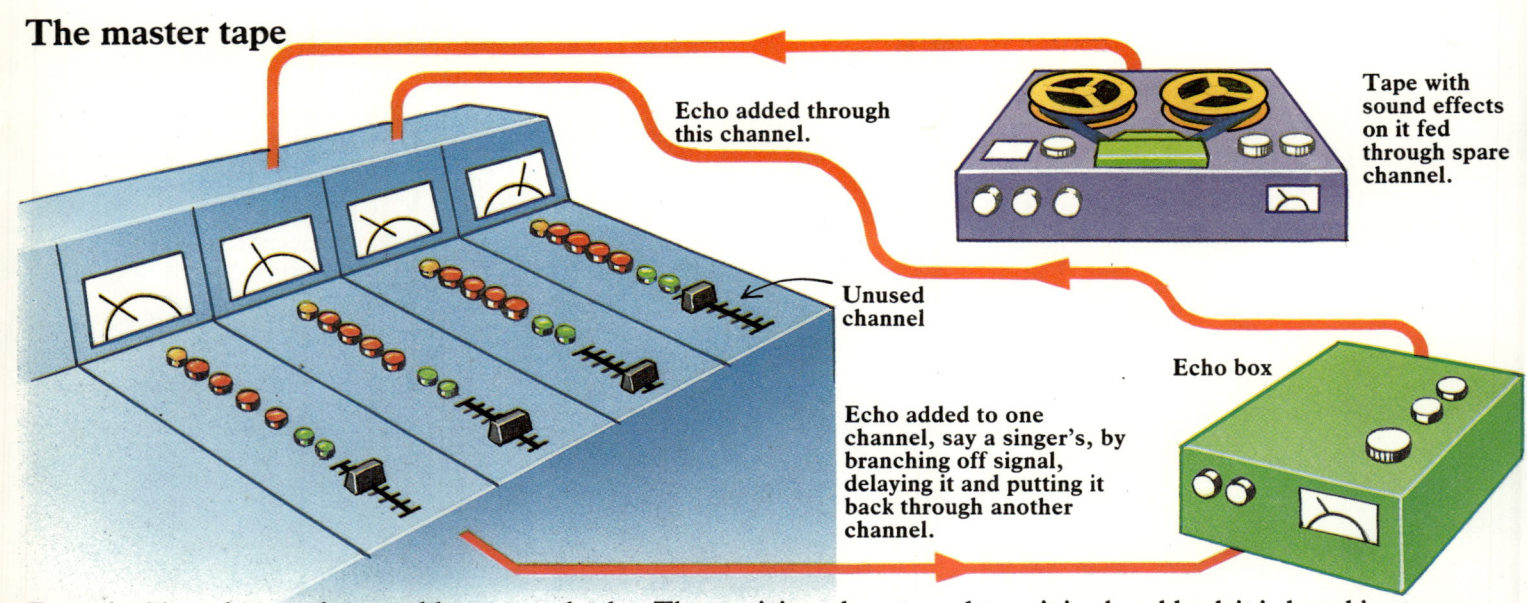

Tape with sound effects on it fed through spare channel.

Unused channel

Echo box

Echo added to one channel, say a singer's, by branching off signal, delaying it and putting it back through another channel.

From the 24-track tape, the sound has to be "mixed down". This is done by feeding the signals back through the desk again from the recorder. The engineer makes alterations to the tone and can add things like sound effects and echo. The musicians do not need to be there. In this diagram you can see how the last few modules of a desk might be used in the mix down.

The sounds are usually put on to a two-track tape eventually, so that when it is played back it is heard in stereo. (The sound from one track comes out of the left speaker, the sound from the other out of the right.) This two-track, stereo tape is ¼in wide and is called the master tape.

Producing stereo sound

When engineers are mixing down, they do not just put half of the 24 tracks on to one track of the stereo tape and the other half on the other track. That would mean the sound of half the instruments came out of one speaker and half out of the other. By using the pan pot, the engineer can make it sound as though the musicians are arranged on stage at a live concert. This diagram shows what happens if the pan pot knob on a module is moved towards the left. If the knob is moved towards the right, the opposite happens.

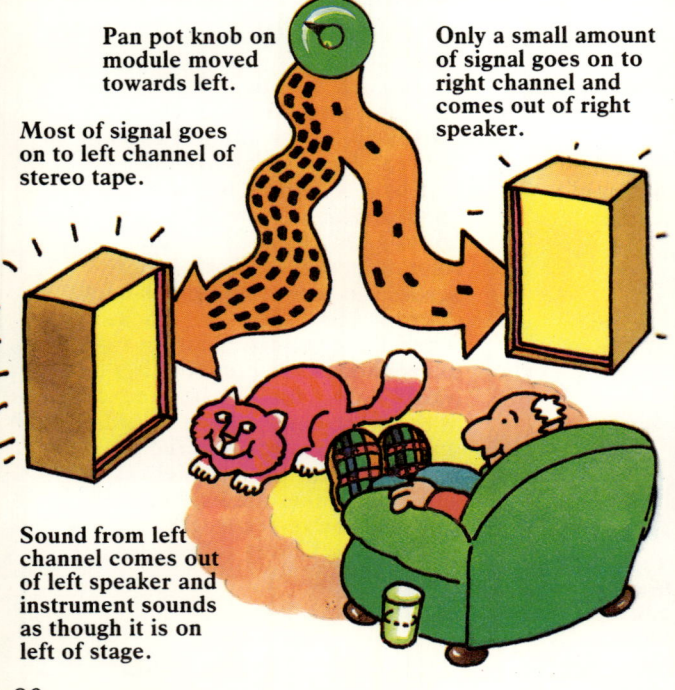

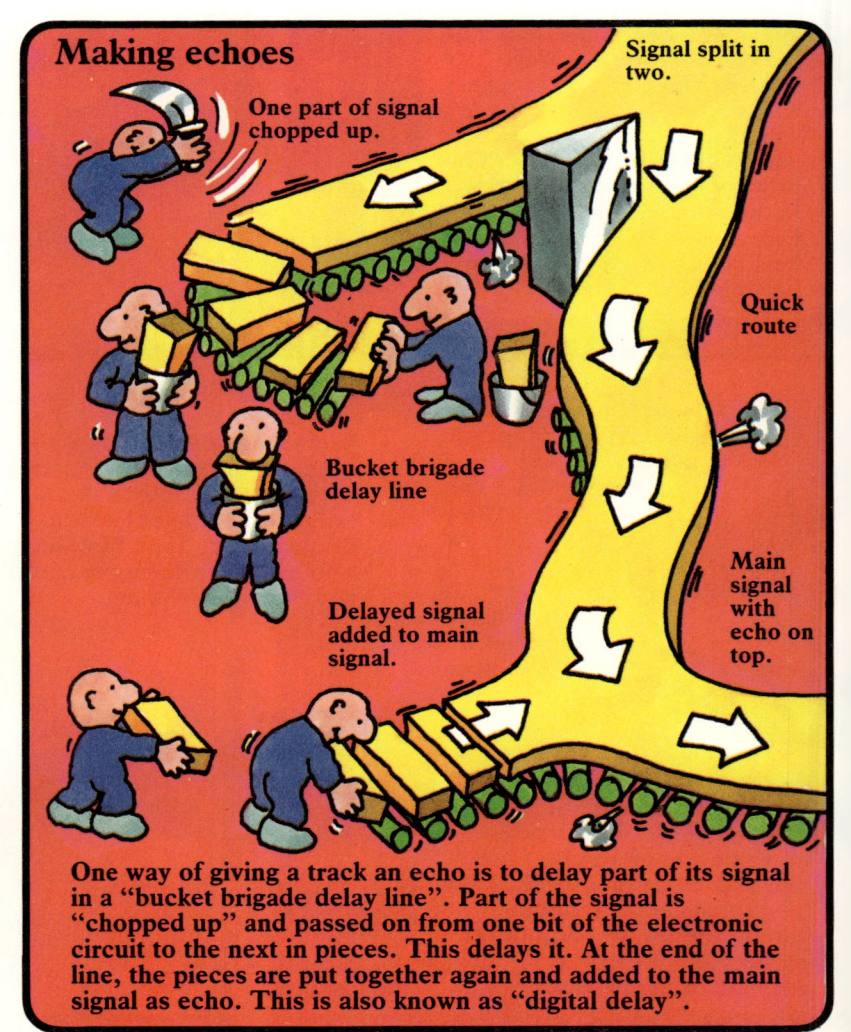

Pan pot knob on module moved towards left.

Most of signal goes on to left channel of stereo tape.

Only a small amount of signal goes on to right channel and comes out of right speaker.

Sound from left channel comes out of left speaker and instrument sounds as though it is on left of stage.

Making echoes

Signal split in two.

One part of signal chopped up.

Quick route

Bucket brigade delay line

Main signal with echo on top.

Delayed signal added to main signal.

One way of giving a track an echo is to delay part of its signal in a "bucket brigade delay line". Part of the signal is "chopped up" and passed on from one bit of the electronic circuit to the next in pieces. This delays it. At the end of the line, the pieces are put together again and added to the main signal as echo. This is also known as "digital delay".

1 Cutting a record

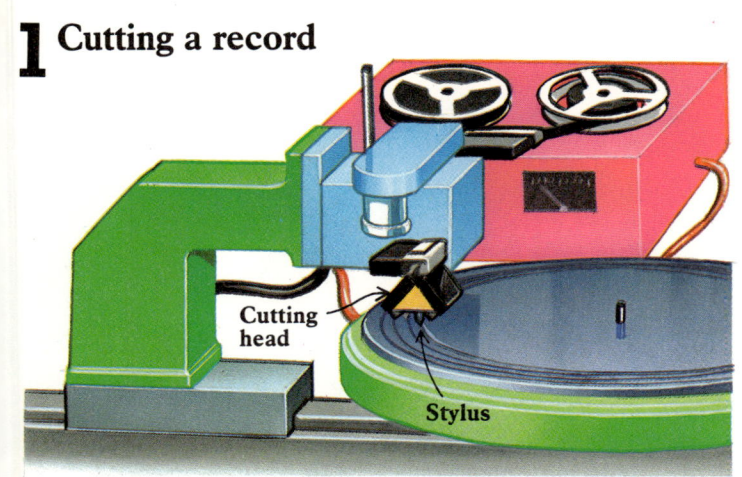

The master tape is played back on a machine which feeds the signals to a "cutting head". The head has a stylus made of diamond. This cuts a groove in a soft plastic disc which is rotated on a turntable. It starts at the outside of the disc and spirals in to the centre.

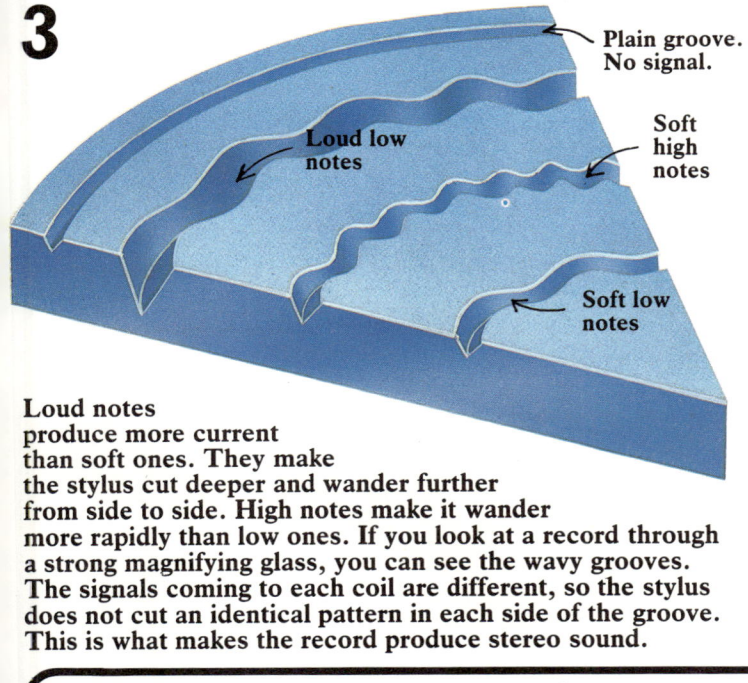

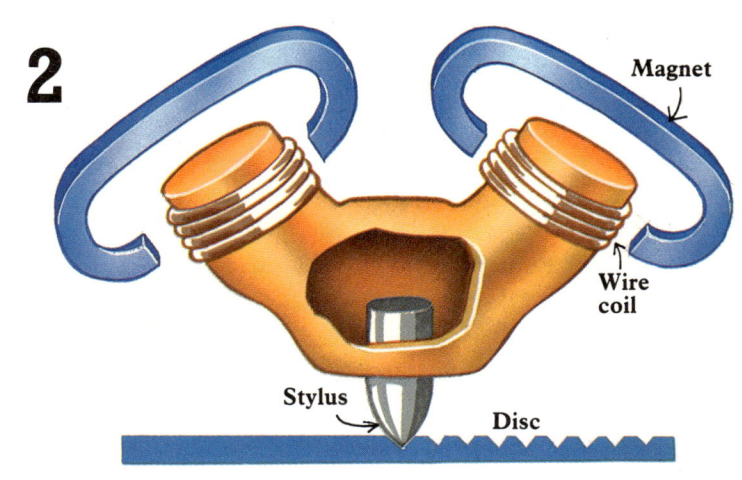

2

The cutting head has two coils of wire attached to the stylus, one for each track of the tape. When the electric signals from the tape reach the coils, the coils vibrate because they are near magnets. This makes the stylus vibrate. The rate of vibrations varies with the current in the coils.

3

Loud notes produce more current than soft ones. They make the stylus cut deeper and wander further from side to side. High notes make it wander more rapidly than low ones. If you look at a record through a strong magnifying glass, you can see the wavy grooves. The signals coming to each coil are different, so the stylus does not cut an identical pattern in each side of the groove. This is what makes the record produce stereo sound.

4

A strong metal "stamper" is made from the disc. It has ridges in place of the grooves. Two stampers (one for each side of the record) are put face to face in a press. They have the record labels on them upside down. One stamper has a ball of PVC material on it. The stampers are heated and brought together. The ridges press into the PVC, "printing" both sides of the record. The record is cooled, trimmed and put in its sleeve.

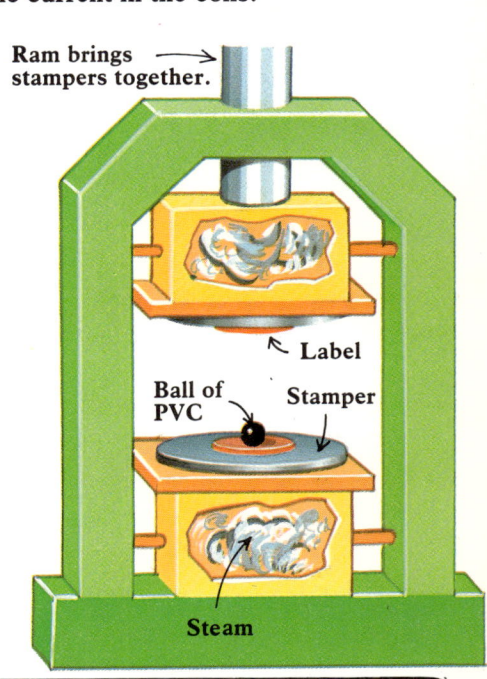

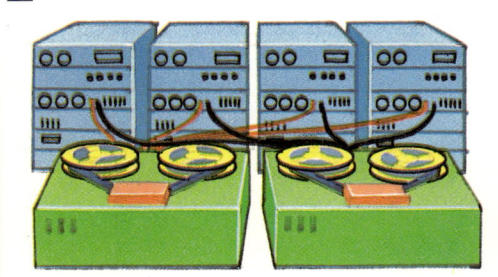

1 Producing cassettes

Cassettes are easier to produce than records. Two master tapes (one for each side of the cassette) are played back at the same time on machines feeding lots of cassette recorders at once. The tapes are speeded up. A cassette which plays for 90 minutes is recorded in 2½ minutes.

2

First side of cassette (tracks 1 and 2) is played to end.

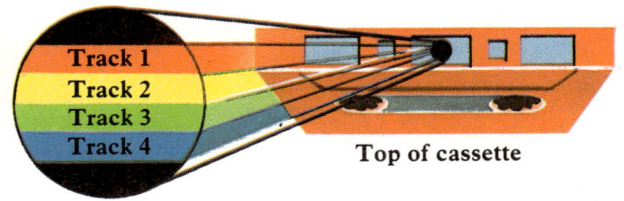

Cassette is turned over for tracks 3 and 4, and starts at end where it just finished. Sound would be back to front if master tape had not been played backwards when recording.

Tape can only be recorded on one face. Cassette tape is ⅛in wide and has four tracks on it. This gives you two "sides" of two-track, stereo sound. One of the master tapes has to be played backwards for the sound to come out the right way round when you turn the cassette over.

81

Audio equipment

These two pages show you some of the different types of audio equipment that are available and how they can be combined. If you are thinking of buying equipment, it is a good idea to plan ahead, as you might be able to build up a good system in stages. Always get any item of equipment demonstrated in the shop before you buy. Remember that the word "hi-fi" is often used instead of "audio" but it really only applies to top quality equipment.

On the right is a typical home hi-fi system of "separates". If you want to build up a system of separate units made by different manufacturers, check with them or the supplier that the units will all work together.

▲ If you want to play records, you will need a turntable. You can get turntables with built-in amplifiers, but these are getting rare and are not usually very good quality.

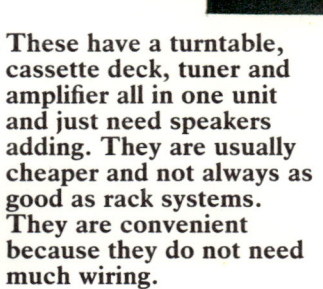

You can use cassette decks for playing tapes and recording, either from another piece of hi-fi equipment or live from a microphone.

Rack systems

These have a separate ▶ turntable, cassette deck, tuner and amplifier made by the same manufacturer and stacked in a rack. There are speakers to match. Most rack systems are good quality and fairly expensive. You can buy the units individually and build up the system gradually.

Music centres

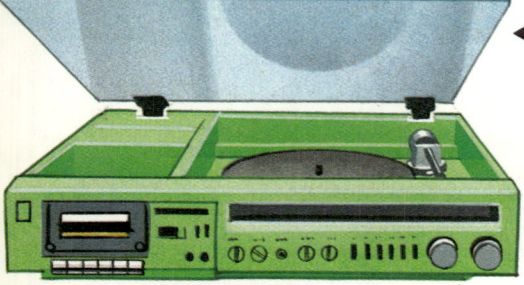

◀ These have a turntable, cassette deck, tuner and amplifier all in one unit and just need speakers adding. They are usually cheaper and not always as good as rack systems. They are convenient because they do not need much wiring.

Mini hi-fi systems

These include a cassette ▶ deck, tuner, amplifier and speakers, and sometimes a turntable, all in a space about the size of a small suitcase. The units can be positioned apart or kept together. Many are portable. Some are about as good as rack systems but they are quite expensive.

Radio/cassette recorders

▲ Portable radio/cassette recorders like this are often quite good quality. The speakers are not far enough apart to give a very good stereo effect but many models can be connected to headphones or separate speakers. One advantage is that you can record from the radio without having to connect two machines. Many of the radios have four wavebands. There is often a socket for a separate microphone. The machines work from batteries or the mains. You can also get portable mono (one track) radio/cassette recorders, with just one speaker. These are cheaper.

Smaller stereo ▶ radio/cassette recorders have headphones instead of speakers. Most work only from batteries. There are often sockets for a second pair of headphones and a separate microphone.

◀ Speakers provide stereo sound. You can sometimes link up three or four speakers to your amplifier for a "surround-sound" effect.

The amplifier boosts the signals from the turntable, cassette deck or tuner before sending them to the speakers. ▼

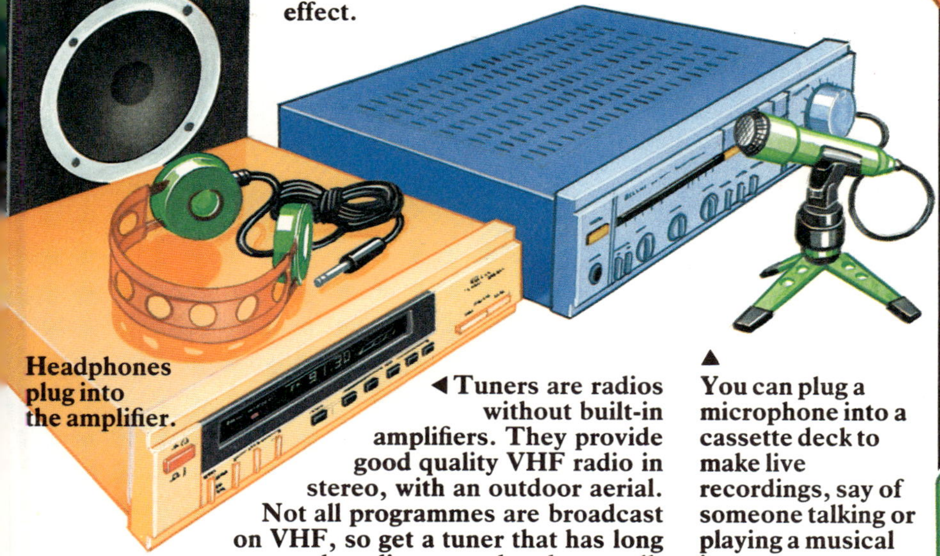

Cassette recorders

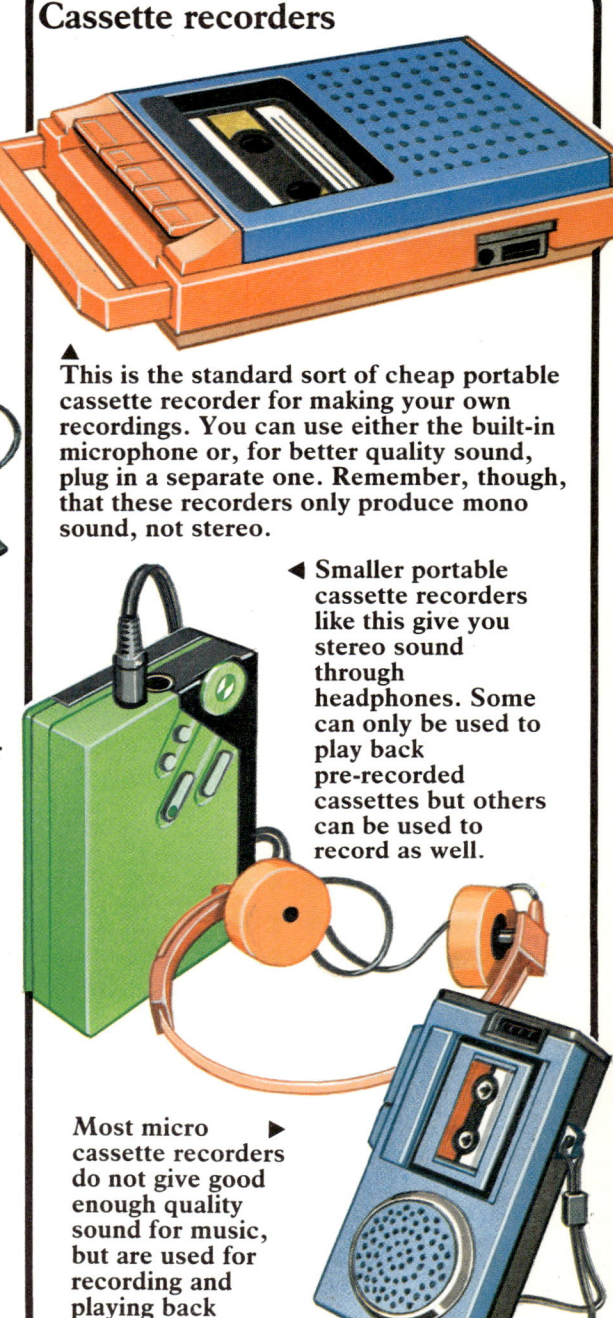

▲ This is the standard sort of cheap portable cassette recorder for making your own recordings. You can use either the built-in microphone or, for better quality sound, plug in a separate one. Remember, though, that these recorders only produce mono sound, not stereo.

Headphones plug into the amplifier.

◀ Tuners are radios without built-in amplifiers. They provide good quality VHF radio in stereo, with an outdoor aerial. Not all programmes are broadcast on VHF, so get a tuner that has long and medium wavebands as well.

▲ You can plug a microphone into a cassette deck to make live recordings, say of someone talking or playing a musical instrument.

◀ Smaller portable cassette recorders like this give you stereo sound through headphones. Some can only be used to play back pre-recorded cassettes but others can be used to record as well.

Radios

Most portable radios do not have more than three wavebands and some only have two. They produce mono sound. They can be run from the mains as well as batteries. It is cheaper to use the mains when possible. ▼

If radio is to be the most important part of your hi-fi system, you might start by buying a "receiver". This is a tuner and amplifier in one unit, and produces good stereo sound through hi-fi speakers. A turntable and cassette deck can be connected up to the amplifier part of the receiver. ▼

Most micro ▶ cassette recorders do not give good enough quality sound for music, but are used for recording and playing back speech.

Car audio equipment

▲ Radio alarm clocks can be set to wake you up to the sound of the radio. Most have three wavebands and produce mono sound. Some have a cassette recorder combined. They run from the mains.

▲ Most micro radios have only two wavebands. They produce mono sound and work from batteries.

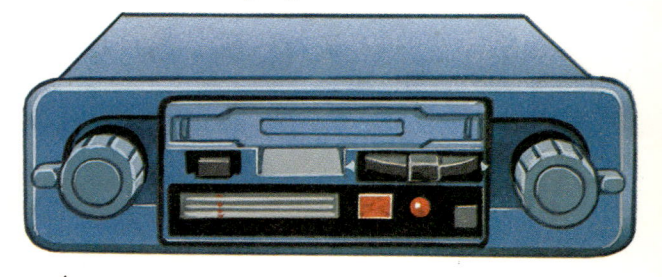

▲ For cars you can get radios and cassette players either combined or separate, often with a clock. Speakers can be fitted to the doors or the back shelf for stereo sound. There are also special mini hi-fi systems for cars.

83

How audio equipment works

Although there are many different types and makes of audio equipment, they all work on the same basic principle. That is, they have to convert sound which is stored on tape or disc back into electric signals and then into sound that you can hear. Here you can find out how the equipment works and pick up some tips and hints for when you are buying equipment.

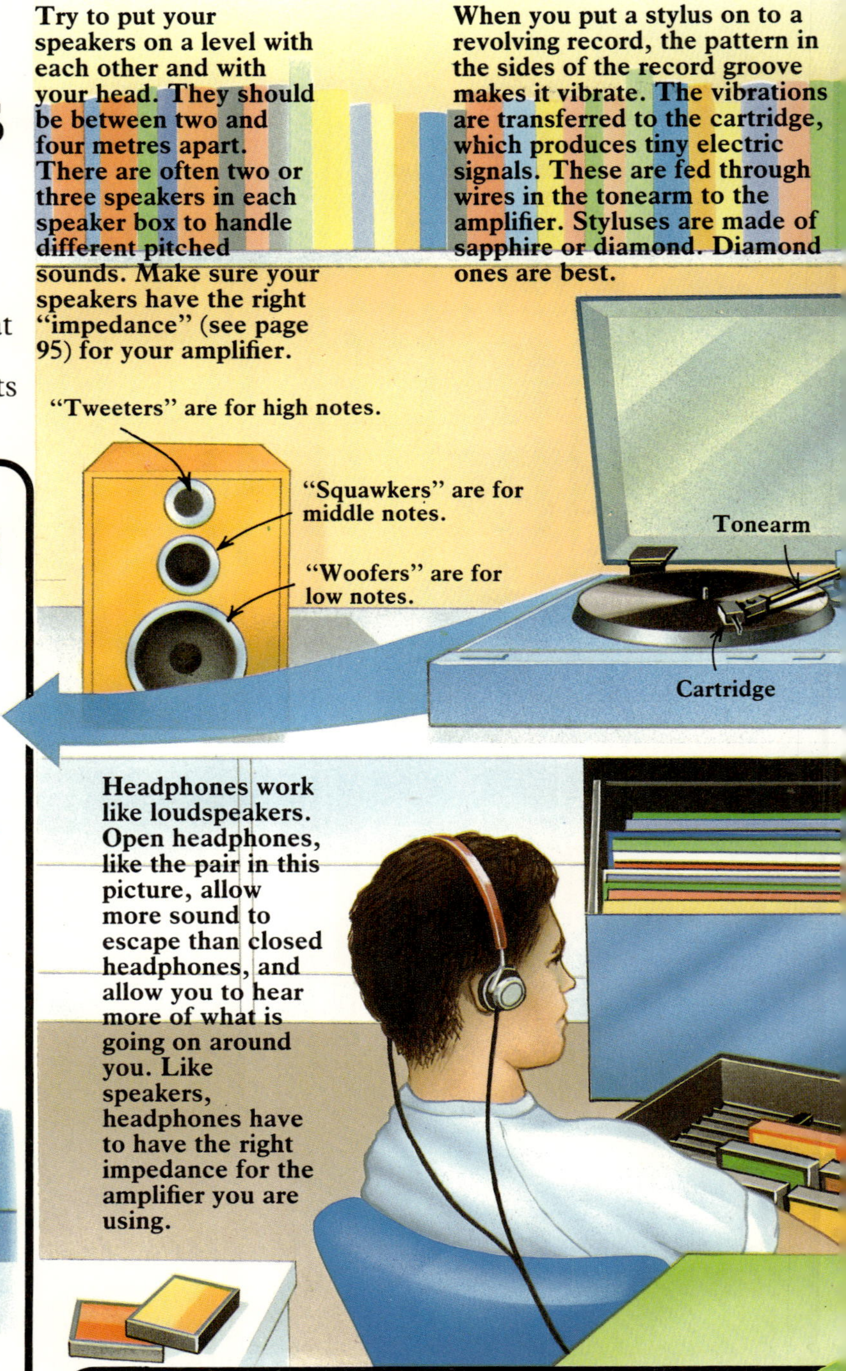

Turntables

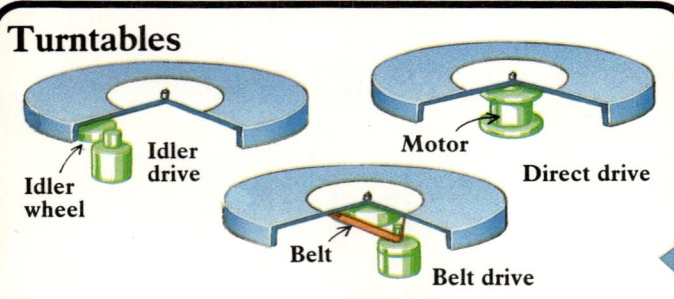

There are three main types of turntable. In the cheapest type, the motor turns a rubber idler wheel, and this makes the turntable go round. Vibrations from the motor are sometimes transferred to the turntable by the idler wheel, producing "rumble" in the speakers. In belt-driven systems, the motor is connected to the turntable by a rubber belt. This cuts down vibrations. The best and most expensive type of turntable is direct drive. Here, the turntable is mounted directly on a special, slow-turning motor.

Tonearms

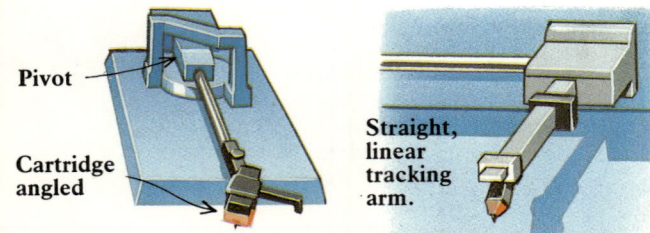

Most tonearms have a pivot at one end and are bent at the other, near the cartridge. This holds the stylus at the correct angle. The best arms are "linear tracking". These are completely straight and are driven along a bar as the record plays. They are very expensive.

Cartridges

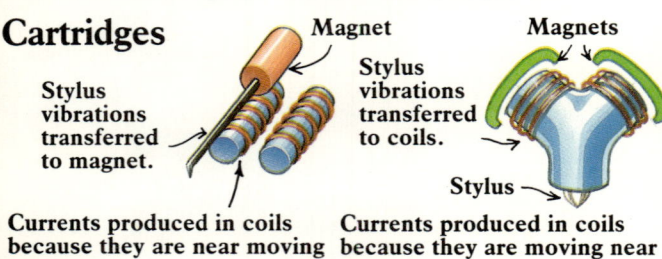

Currents produced in coils because they are near moving magnet.

Currents produced in coils because they are moving near magnets.

There are two main sorts of cartridge: moving magnet and moving coil. Moving coil types work on the same principle as the disc cutting cartridge on page 81. Make sure your cartridge is suitable for your amplifier.

Try to put your speakers on a level with each other and with your head. They should be between two and four metres apart. There are often two or three speakers in each speaker box to handle different pitched sounds. Make sure your speakers have the right "impedance" (see page 95) for your amplifier.

"Tweeters" are for high notes.
"Squawkers" are for middle notes.
"Woofers" are for low notes.

When you put a stylus on to a revolving record, the pattern in the sides of the record groove makes it vibrate. The vibrations are transferred to the cartridge, which produces tiny electric signals. These are fed through wires in the tonearm to the amplifier. Styluses are made of sapphire or diamond. Diamond ones are best.

Headphones work like loudspeakers. Open headphones, like the pair in this picture, allow more sound to escape than closed headphones, and allow you to hear more of what is going on around you. Like speakers, headphones have to have the right impedance for the amplifier you are using.

Cassette decks

Erase head
Recording/playback head
Pinch wheel holds tape against capstan.
Capstan

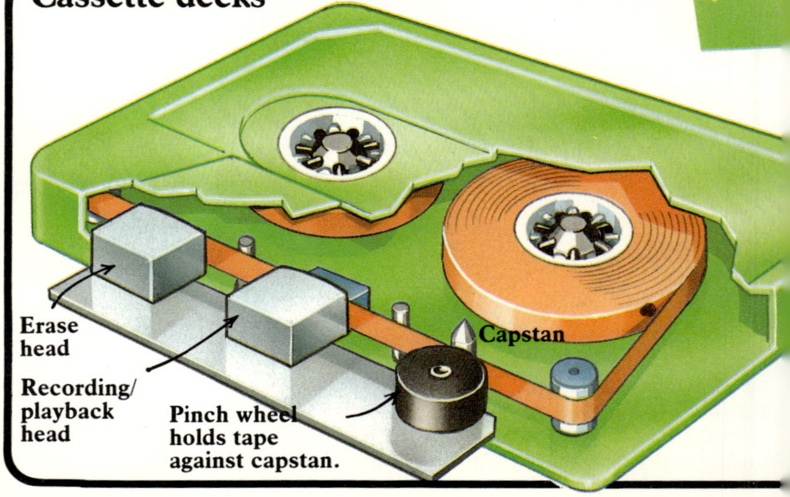

Some people think records produce better quality sound than cassettes. On the other hand, they take up more room and are more easily damaged. Digital records are better than ordinary ones. You can find out more about these on page 94. Always keep records in their sleeves and stack them side by side. Clean them with special cloths from record shops.

Amplifiers

Signals boosted.

Tone and balance controlled.

Signals boosted again.

All electric sound signals have to be amplified to make them strong enough to drive loudspeakers. The amplifier picks up tiny signals from whichever piece of equipment, or "source", you switch on, and boosts them by transistors. It sets the tone and balance of the sounds. Then it boosts the signals about a million times more. It is important not to buy an amplifier too powerful for the system's speakers. Sometimes, instead of being "integrated", an amplifier is split into two parts: a preamplifier and a power amplifier.

There are many different types of cassette tape. Chrome dioxide (CrO_2) and metal are the best. Cassettes are labelled C10, C30, C60, C90 or C120. The figures show the total running time (both sides) in minutes. C120s are made of thinner plastic than the others so enough tape can be squeezed into the cassette. This means they break more easily.

Keep cassettes in a cool, dry place, away from electrical equipment, which can erase recordings. This includes vacuum cleaners, TV sets and even speakers. Playing a head-cleaning cassette occasionally helps to preserve the quality of the recordings.

You will come across two main types of microphone: dynamic or moving coil (see page 67) and capacitor or condenser. These can both be either omni-directional, which means they pick up sound from all around, or uni-directional (cardioid), which means they pick up sound mainly from the front.

Loudspeakers

Electric signals come from the amplifier to a wire coil attached to the speaker cone. This electric current makes the coil vibrate, because it is near a magnet. This then makes the cone vibrate, producing sound waves in the air. Speakers work like moving coil microphones in reverse (see page 67).

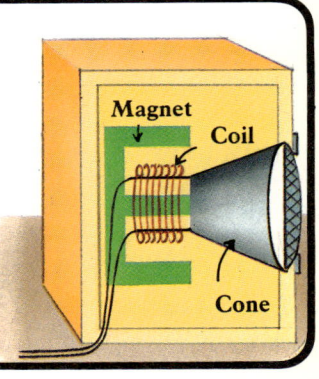

Magnet *Coil* *Cone*

Tape is pulled along by a capstan, driven by an electric motor. Sound is stored on tape in a magnetic pattern. When a recorded tape passes across the playback head, it magnetizes two iron blocks in the head (one for each stereo track). This produces electric currents in the wires wrapped round each block. The currents vary according to the magnetic pattern on the tape. They go to the amplifier. (To find out how sound gets on to tape when you make a recording, see page 78.)

Most cassette decks have two heads, one for playback and recording, the other for erasing. When you press the erase button, a supersonic current, produced in a special circuit, destroys the magnetic pattern on the tape, leaving it ready for a new recording.

Reel-to-reel recorders

These are expensive and are used by recording enthusiasts. There are not many pre-recorded tapes for them. They use wider tape than cassette decks (¼in) and can be played at faster speeds. This improves the quality of the sound. You can edit by cutting as well as erasing (see page 79).

85

The controls on audio equipment

These two pages show you what the different controls on audio equipment are for and explain some technical audio words. The controls are shown here on a rack system but many are very similar on individual or smaller, portable pieces of equipment. Not all equipment has all these controls. On pages 94-95 there are some more technical words that you will come across, particularly in makers' specifications for equipment.

SPEED SELECT. Records have to be spun at the same speed as when they were being cut. These controls make the turntable revolve at 33 r.p.m. (revolutions per minute) for LPs (long play) and 45 r.p.m. for EPs (extended play).

RECORD/PLAYBACK. You press these controls to record, play back a tape, re-wind, move the tape forward fast, pause while recording, and to stop the tape and open the door so the cassette can be taken out.

POWER INDICATORS. These show the power output of the amplifier. Here, a lot of power is going to the left speaker and not much to the right. When an amplifier's output reaches the top of its power range, the sound often gets distorted.

SOURCE OR INPUT. This control selects the signals from whichever piece of equipment you want to play. This amplifier can handle signals from a turntable with either a moving magnet or moving coil cartridge, from a tuner, or from either of two cassette decks or reel-to-reel recorders. The auxiliary position means you can link up another piece of equipment to your system, perhaps a friend's cassette deck for recording from their machine to yours, or an electronic musical instrument.

SIGNAL STRENGTH INDICATOR. This shows whether you are tuned to a station exactly and whether you are receiving signals as clearly as you should.

STROBOSCOPE. Some turntables have patterns of bars round the edge and a neon light which shines on them. When the turntable speed is correct, the bars look stationary under the light.

QUARTZ LOCK. Quartz-locked turntables have a special circuit for keeping their speed accurate. This cuts down "wow" (distorted sound caused by slow fluctuations in speed) and "flutter" (distortion caused by rapid fluctuations in speed). The phrase "phase-locked loop" (PLL) also applies to these turntables.

POWER. The on/off control supplies mains electricity to power the equipment.

TWO MOTOR. Some cassette decks have one motor to drive the capstan and another to drive the spools. This helps keep the tape speed constant and so cuts down wow and flutter (see above).

PHONES. Check that your headphones' "impedance" is suitable for the deck.

TAPE COUNTER. This shows how far along the tape you are, so you can make a note of where to find a particular track. Pressing the re-set button puts the counter to zero.

PHONES. You can monitor what you are recording through headphones.

TAPE TO TAPE. Press this control to record from one tape machine to another.

PRE-SET STATIONS. You can often pre-set several stations, usually on the VHF waveband, and keep them stored in the tuner's memory. Press the VHF control and turn the tuning knob until the exact frequency you want shows up on the display. Then press one of the memory buttons. By pressing the same memory button again, you recall the station, ready tuned.

QUARTZ SYNTHESIZER. This is a special circuit for very precise tuning.

DIGITAL TUNING DISPLAY. This shows precisely the frequency you are tuned to.

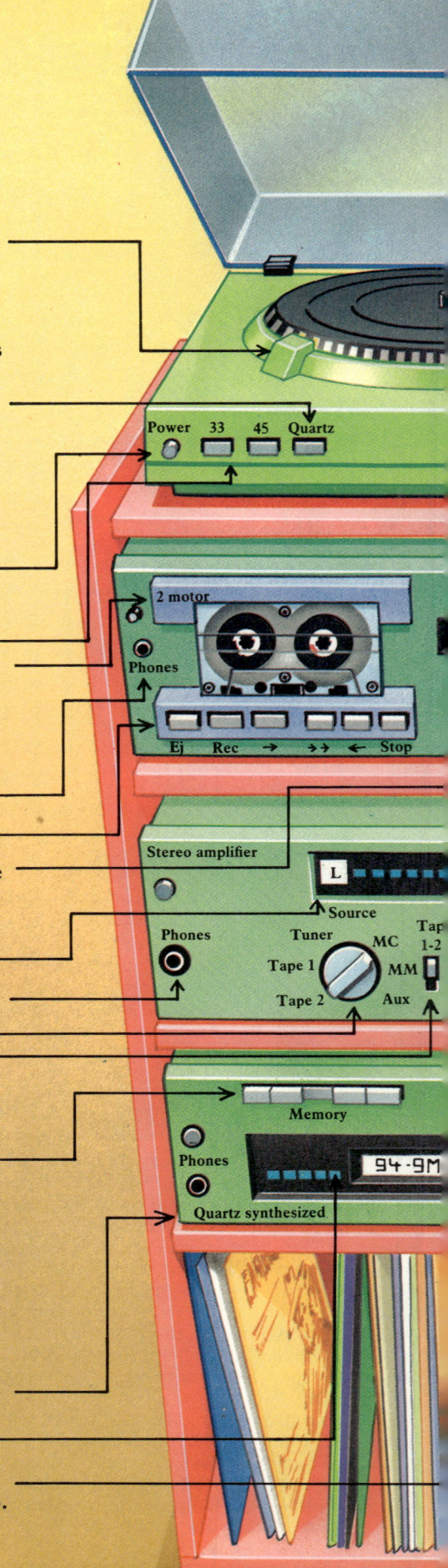

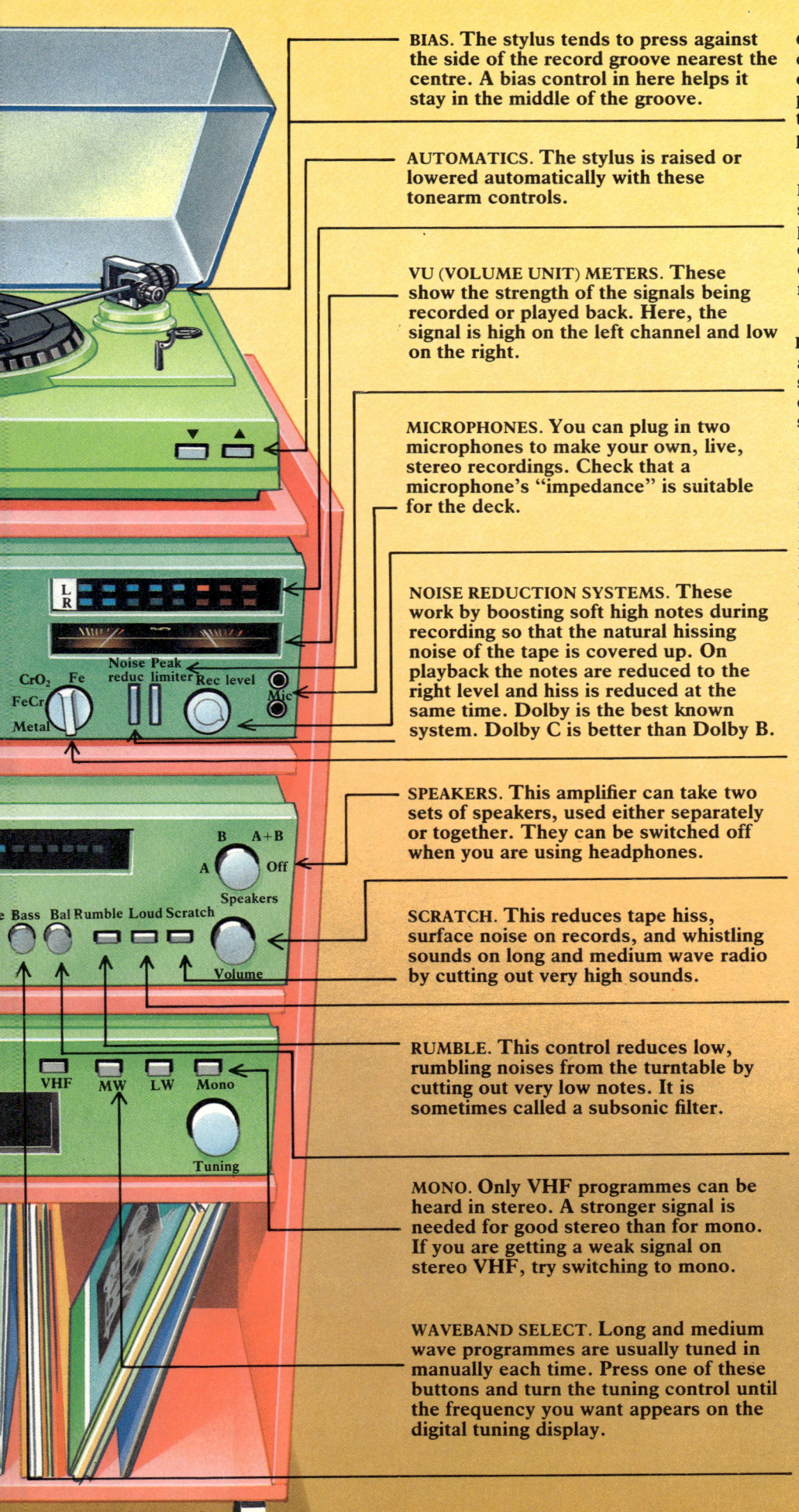

BIAS. The stylus tends to press against the side of the record groove nearest the centre. A bias control in here helps it stay in the middle of the groove.

AUTOMATICS. The stylus is raised or lowered automatically with these tonearm controls.

VU (VOLUME UNIT) METERS. These show the strength of the signals being recorded or played back. Here, the signal is high on the left channel and low on the right.

MICROPHONES. You can plug in two microphones to make your own, live, stereo recordings. Check that a microphone's "impedance" is suitable for the deck.

NOISE REDUCTION SYSTEMS. These work by boosting soft high notes during recording so that the natural hissing noise of the tape is covered up. On playback the notes are reduced to the right level and hiss is reduced at the same time. Dolby is the best known system. Dolby C is better than Dolby B.

SPEAKERS. This amplifier can take two sets of speakers, used either separately or together. They can be switched off when you are using headphones.

SCRATCH. This reduces tape hiss, surface noise on records, and whistling sounds on long and medium wave radio by cutting out very high sounds.

RUMBLE. This control reduces low, rumbling noises from the turntable by cutting out very low notes. It is sometimes called a subsonic filter.

MONO. Only VHF programmes can be heard in stereo. A stronger signal is needed for good stereo than for mono. If you are getting a weak signal on stereo VHF, try switching to mono.

WAVEBAND SELECT. Long and medium wave programmes are usually tuned in manually each time. Press one of these buttons and turn the tuning control until the frequency you want appears on the digital tuning display.

COUNTERBALANCE. This counterbalances the tonearm and cartridge. It has to be positioned very precisely so that the stylus sits lightly on the record with just the right amount of pressure or "tracking force".

PEAK INDICATORS. These show the strength of signals being recorded or played back. They use "light emitting diodes" (LEDs), which respond to changes in signal level faster than VU meters do.

PEAK LIMITER. During recording, this automatically reduces very high level signals, produced, for example, by a clash of cymbals. This helps prevent the sound getting distorted.

RECORDING LEVELS. These controls increase or decrease the strength of the signals being recorded. They should be set so that the needles on the VU meters just reach the red sections in the loudest passages of music. If the levels are too high, the sound will be distorted on playback. If they are too low, you will hear hiss.

EQUALIZATION. When cassettes are being made, the balance between the different pitched notes is artificially altered to get the best results on tape. The balance has to be restored, or equalized, on playback to suit the type of tape. This deck has special circuits for ferro, chrome dioxide (CrO_2), ferro-chrome and metal tapes.

VOLUME. This control alters the amount of signal fed to the speakers or headphones.

LOUDNESS. When the volume is turned down low, say late at night, it can be difficult to hear very high and very low notes. A loudness control boosts the volume of these notes slightly to compensate.

BALANCE. This control affects the relative amounts of sound coming from each speaker. When it is in the central position, equal amounts come from both speakers.

TREBLE AND BASS. You can get the tone of the sounds as you like them with these controls. The treble increases or decreases the volume of the high notes, the bass does the same for the low notes. "Graphic equalizers" are special pieces of equipment which give you more precise control of tone than an ordinary amplifier does.

87

Make your own radio

These instructions are for building a simple, medium wave radio. You will be able to buy the components you need from an electronics components shop or, if there is no shop near you, by post from an electronics supplier. Look for suppliers' advertisements in electronics hobby magazines. You should be able to buy all the other things from a general electrical or hardware shop.

It is important to follow each step of the instructions very carefully, as the tiniest mistake will probably mean the radio does not work. The number of stations you pick up on your radio will partly depend on where you live. At the bottom of page 90 are some tips on what to do if you live in an area of poor reception.

Components
(You will need one of each of the following.)

Resistors
($\frac{1}{8}$ or $\frac{1}{4}$ watt, preferably with 5% tolerance, though 2% will do.)

100Ω, 270Ω, 680Ω, 3.9kΩ, 15kΩ, 39kΩ, 100kΩ (Ω means *ohm*. kΩ means 1,000 *ohms*.)

Variable resistor
25kΩ potentiometer for use as volume control.

Capacitors (wire ended type)
10nF, 100nF (nF means *nanofarad*.)

Electrolytic capacitors
(Axial lead type, 9 volt. If the voltage is any higher, you may find the capacitors are too big for this circuit layout.)

4.7μF, 22μF, 47μF (μF means *microfarad*.)

Variable capacitor
500pF (*picafarad*), single gang, without trimmers, for use as tuner.

Diodes (two of these) IN4148

Transistor BC107

Integrated circuit
ZN414

Audio transformer
With impedance ratio of 1 or 1.2kΩ: 3.2 or 8Ω, e.g. LT700.

3 or 8Ω loudspeaker (that is, to match your audio transformer), about 80mm in diameter.

Single pole, on/off, slide switch.

Veroboard with copper strips, 0.1in spacing, at least 30 tracks x 32 holes.

2 metres of enamelled copper wire, s.w.g. 30.

Ferrite rod, 9 or 10mm in diameter, 10cm long.

Other things you will need

Soldering iron, cored solder, wire cutters, sharp knife, small pliers, drill bit (about 4mm), 9 volt battery, sponge, clear sticky tape, needle and thread, 2 or 3 metres of solid conductor insulated "hook-up" wire, thin enough to go through the holes in the Veroboard when stripped. Twin-wire cable can be split in two.

Note. If you are not used to soldering or working with Veroboard, it is a good idea to buy more of the cheapest components (resistors) than you need and an extra piece of Veroboard, so you can practise.

How to solder

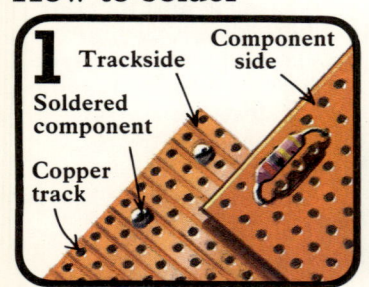

1. When components are soldered on to Veroboard, electric current from a battery will flow to them along the copper strips on the back, or trackside, of the board.

2. You need a soldering iron, some cored solder and a damp sponge. Plug the iron in and wait for it to heat up. *Prop the iron up carefully so it doesn't burn anything and don't touch the bit.*

3. Always make sure you put the legs of the components through the right holes in the Veroboard (see next page). Bend them away from each other slightly at the back to hold them in place.

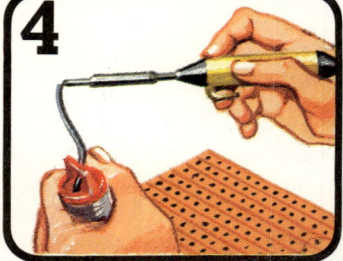

4. Wipe the hot bit on the damp sponge to clean off any old solder. Then, touch the bit with the end of the solder wire. The solder will melt immediately and a drop of solder will cling to the bit.

5. Then, both at the same time, put the bit and the solder wire on the track, up against the component's leg. Leave them there for only a few seconds until the solder melts and joins the leg to the track.

6. Let the "joint" cool and then trim the leg about 1mm from the solder with wire cutters. Hold the leg as you cut to stop it flying into the air. Be careful. Bits of flying metal can be very dangerous.

7. Run iron along groove. Joints should be smooth and shiny and there must not be any solder between tracks or the current will flow across them. To remove solder, run the hot iron along the groove between the tracks.

8. *Remember to unplug the iron when you have finished.* If you solder a component in the wrong place, you can remove it by melting the joint. Put the hot iron on the joint and gradually lift off the solder, wiping the bit on the damp sponge.

1 Building the radio

Hold knife firmly against ruler.

If your Veroboard is bigger than you need, you can cut it down. Score along a row of holes on the trackside several times using a sharp knife and metal ruler. Then break the board. Leave a few more tracks and holes than you need in case you don't break the board quite straight.

2

Make a grid to help you find where to put the components. Put the board on a piece of paper component side up, with the tracks running horizontal. Draw round the board. Make marks for the holes along the top and for the tracks down the side. Number the holes 1-32 and the tracks A-DD, as shown above.

3

Broken track

You need to "break tracks" in places. Turn the drill bit in these holes until the copper either side is removed: A14, A18, F26, G12, H14, H20, I18, M21, N11, N21, O21, R22, S11, AA22 and BB22. When you find a hole, try sticking a piece of wire through so you don't lose it again when you turn the board over.

4

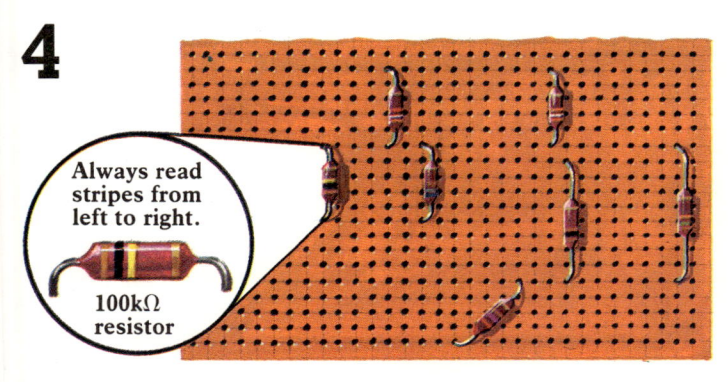

Always read stripes from left to right.

100kΩ resistor

Solder the resistors to the board first. You can tell which is which by their coloured stripes. (Ignore the gold or red stripe round one end.) It doesn't matter which way round you put the resistors. The brown, black, yellow striped resistor (100kΩ) goes in holes I9 and N9. The blue, grey, brown (680Ω) in I15, N15. Orange, white, red (3.9kΩ) in N13, S13. Orange, white, orange (39kΩ) in N23, S23. Red, violet, brown (270Ω) in A17, E21. Brown, green, orange (15kΩ) in E31, N31. Brown, black, brown (100Ω) in E24, M24.

5

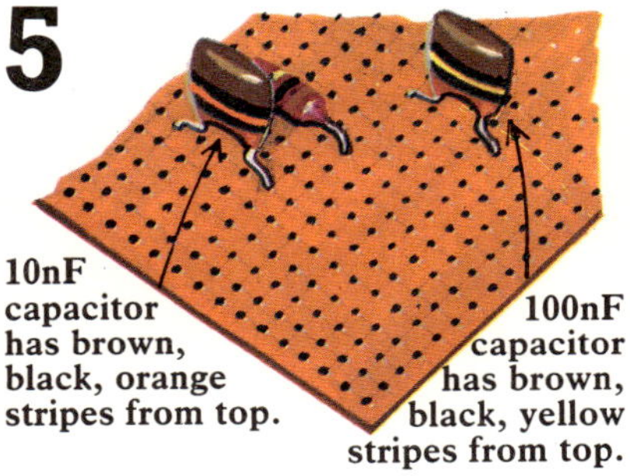

10nF capacitor has brown, black, orange stripes from top.

100nF capacitor has brown, black, yellow stripes from top.

Next, put the 10nF capacitor in J6 and N6, and the 100nF capacitor in E13 and I13. 10nF is sometimes printed on capacitors as 0.01μF and 100nF as 0.1μF. Or, your capacitors may have coloured stripes like the ones shown in the diagram above.

6

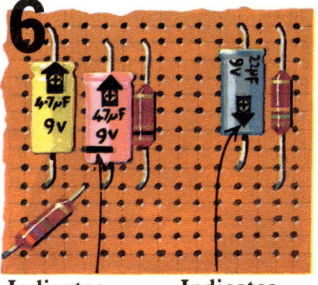

Indicates negative end   Indicates positive end

Electrolytic capacitors have to go a certain way round. Put the 22μF capacitor in F29, N29 with the positive end at F. The 47μF goes in E22, M22 with the positive end at M. The 4.7μF goes in E19, N19 with the positive end at N.

7

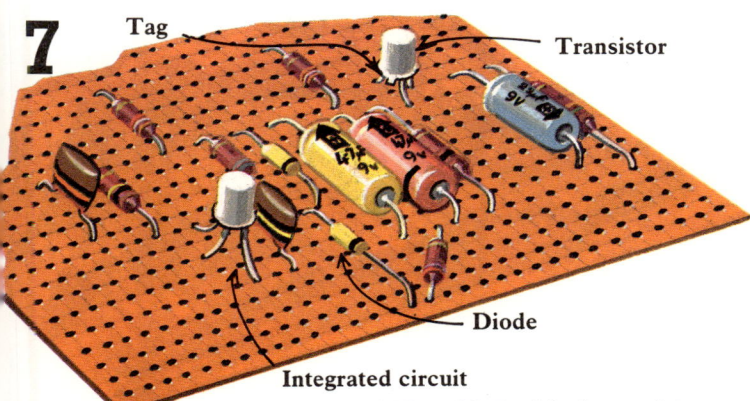

Tag   Transistor

Diode

Integrated circuit

Put the diodes in A16, H16 and H17, N17 with the positive ends (marked with a ring) towards A and H. Don't overheat the diodes, transistor or integrated circuit or you may damage them. The transistor goes in M26, N27, O26, with the leg nearest the metal tag in M26. The integrated circuit goes in E11, G10, I11, with the leg nearest the tag in I11. Leave both the transistor and integrated circuit sticking up about 1cm from the board.

8

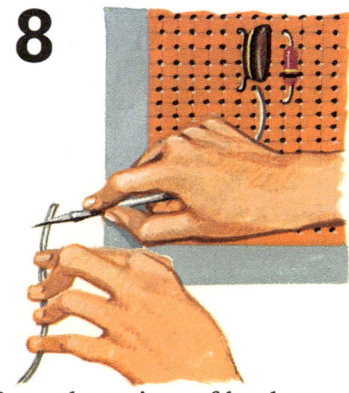

Cut a short piece of hook-up wire. Using a sharp knife, carefully strip about 1cm of the plastic casing off each end. Put one end through hole E7, the other through J7 and solder. Solder another short wire link from O28 to R28.

9

Cut off this wire.

BB24

R25

V21

S21

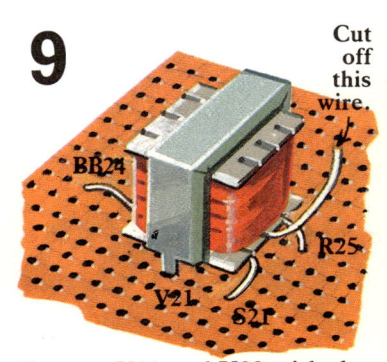

Enlarge V21 and V29 with the drill bit, so the transformer's legs will go through. Put the wires in R25, S21, AA26 and BB24, as shown above. Make sure the transformer is the right way round. Cut off the extra wire. Bend the legs flat on the trackside. Solder them and the wires.

89

MAKE YOUR OWN RADIO

10 Put rod on unglued part of paper and roll up. 35mm. Glue.

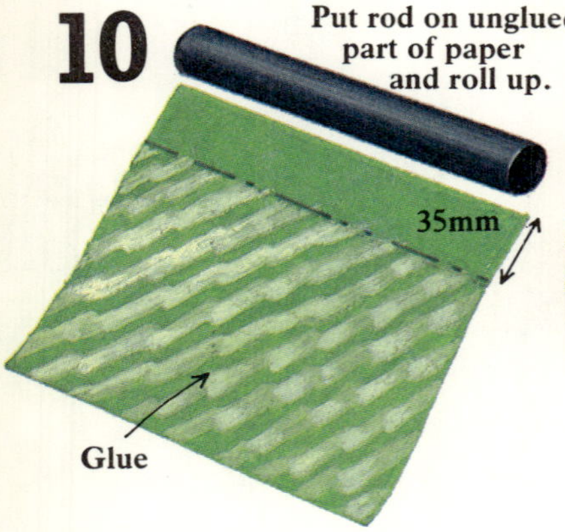

Now make the aerial. If your ferrite rod is too long, cut a groove round it with a hack-saw, 10cm from one end. Then break it. Make a tube for the rod out of a thick piece of paper, as shown above. Put sticky tape round the ends of the tube, so the rod doesn't fall out.

11 Strands of wire must lie side by side.

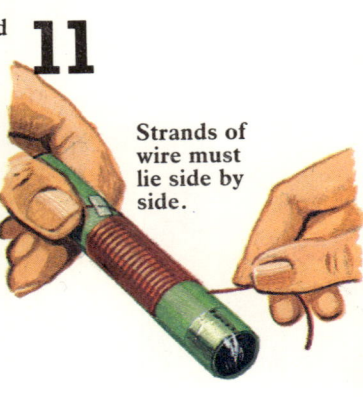

Starting 4cm from one end of the tube and leaving about 5cm of wire free, wind enamelled copper wire neatly round the tube 36 times. Tape the end of the wire to the tube to hold it in place as you wind. Leave 5cm free at the other end. Cover all the wire with clear sticky tape.

12

Scrape both ends of the wire with a knife to remove the coating. Touch the soldering iron with solder and quickly stroke the ends to give them a coating of solder. Then solder them in G5 and N5. Don't stretch the wires tight. Fasten the aerial to the board with a needle and thread.

13

Cut nine pieces of hook-up wire about 15cm long. Strip the casing off the ends and solder one end of each piece in these holes: E14, E30, F28, G1, I14, N1, S31, AA30, BB30.

14 Twist wires together to keep them neat.

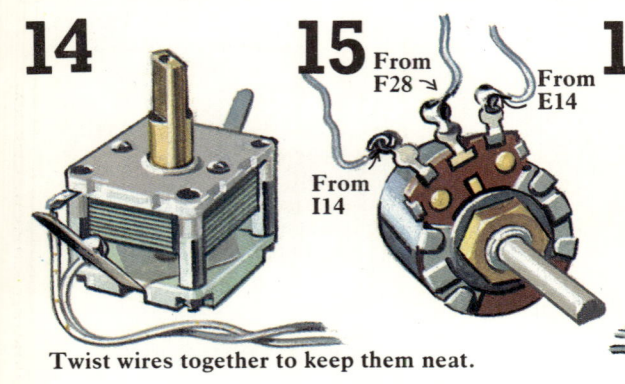

Solder the wire from N1 to one of the long legs on the tuner and the wire from G1 to the short leg.

15 From F28. From E14. From I14.

Solder the wires from E14, I14 and F28 to the volume control as shown above. Hook wires through holes to solder.

16

Solder the wires from AA30 and BB30 to the speaker tags. It does not matter which way round they go.

17 Terminals marked on battery. From E30.

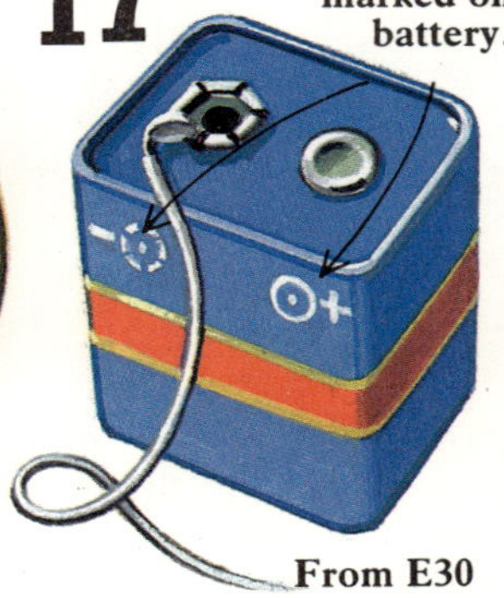

Solder the wire from E30 to the negative battery terminal. Make quite sure you have the correct terminal.

18 Up for off. Ignore top tag. From S31. Wire link from positive battery terminal to bottom tag.

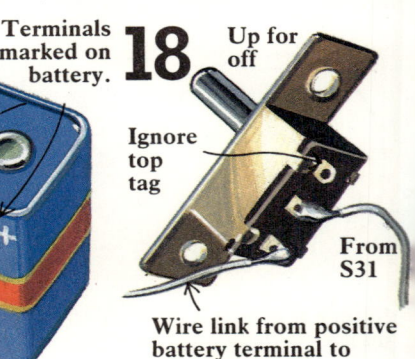

Cut another length of hook-up wire and strip the ends. Connect the switch to the board and battery as above.

Checking the radio

Before you switch on, check all the components on the board to make sure they are the right ones, that they are in the right holes and are the right way round. Check you have broken tracks in the right places and that the wires to the other components are connected correctly. Check again that the battery is the right way round. If you get it wrong, not only will you have to correct it, you may have to replace the transistor, diodes and integrated circuit with new ones, as they could be ruined.

Now, turn up the volume and switch on. Turn the tuner slowly up and down the waveband to see how many stations you can get. If the sound is not very good, try moving the radio around to get better reception.

If the radio doesn't work, switch off and check everything again. Examine all the joints very carefully to make sure they are firm. Try waggling the components or giving them a tug. Make sure there is no solder between tracks and that the components' legs are not touching each other on the component side of the board. Try another battery, in case yours is flat.

If you still can't find anything wrong, ask someone else to check the board for you. They may spot something you didn't. If, after all that, you still can't get the radio working, send it to us and we will try and find out what is wrong. Wrap it up very carefully and send it (with enough stamps to pay for the return postage) to: Electronics Adviser, Usborne Publishing Ltd, 20 Garrick Street, London WC2E 9BJ, England.

Expert's tips

If you live near a strong transmitter, you may only pick up that station. If this happens, take out the resistor at A17, E21 and replace it with a piece of wire.

If you live in an area of weak reception, try rigging up an extra aerial with a few metres of wire. First, put an extra 10nF capacitor at C6, G6 and break the track at C8. Then, solder one end of the aerial wire in C1 and put the other end somewhere high up, for example, over a curtain rail.

If you want, you can box your radio. Make holes for the switch, speaker, and volume and tuning controls, and mount them in position, re-soldering as necessary. Keep the tuner and its wires away from the others. Pad the radio well.

CB and two-way radio

One of the most common uses of radio is for two-way communication. This is especially useful for groups of people who need to communicate while on the move, such as the fire brigade, police, ambulance drivers, pilots, ships' radio officers, taxi drivers and astronauts. Different radio frequencies are allocated to the different users of two-way radio, usually on the VHF or UHF (ultra-high frequency) wavebands. The type of transmitters and receivers used depends on the waveband and the distance the messages have to travel. A "transceiver" is a transmitter and receiver in one unit.

Citizens' band

Anyone can become a "CBer" and talk to other people by radio. All you need is a licence from the Post Office and a transceiver, or rig. The first CB rigs were used in lorries or cars but you can also buy base rigs to use at home, or small pocket-size sets. They are all easy to operate. The signals will travel only a short distance in built-up or hilly areas and up to about 30km in open country. CBers are supposed to keep their conversations short to stop their radio channels getting jammed and they often talk in a number code. If you want to find out more, look in the CB magazines to find out where your nearest club is.

Amateur radio

If you are interested in radio technology, you might like to become an amateur radio "ham" and talk to people all over the world. To do this, you have to be at least 14, hold a licence and pass exams to show you understand the equipment and know how to operate it. A lot of amateurs build at least part of their equipment themselves. It is possible to transmit from a car as well as from home. As well as talking normally, amateurs communicate in Morse code and, for exchanging technical information, in an internationally understood code of three letters. Each radio ham has a "call sign" of letters and numbers, which identifies the individual and the country they are speaking from. They often send "QSL" cards to each other, confirming their conversations. It is illegal to use amateur radio for business or propaganda purposes. To find out more, write to your national radio society, enclosing an s.a.e. They will tell you where the nearest local branch is. (The address in Britain is: Radio Society of Great Britain, 35 Doughty Street, London WC1N 2AE.)

If you want to listen in to amateurs' conversations but don't want to make transmissions, you can use a high quality short wave receiver. You do not need a licence just for listening.

Car telephones

Conversations go from the car to the telephone exchange and back again by radio. The link from the exchange to the other person in the call is made on the ordinary telephone network.

Telephone

Transmitting and receiving aerial.

Transceiver

Different uses of radio and audio

Radio and, less often, audio are used in many ways in addition to the ones explained so far in this book. As well as carrying sound, pulses of radio waves can be used to control things and to find out information about things. Here are some examples of the different uses.

Television

Both the pictures and sound for television programmes are transmitted on radio waves. The TV cameras convert light into electronic signals for the pictures. Then, both these and the sound signals are broadcast on the VHF or UHF wavebands.

Dish aerial

Receiver

Radio waves hit dish and are reflected on to receiver.

Aerial can be steered to face in different directions.

Astronomy

Many objects in space give out radio waves instead of, or as well as, light. Astronomers can study them using radio telescopes like the ones on the left. The telescopes act like giant radio receivers. They pick up the signals from space, which can then be recorded on tape and fed to computers for analysis.

Radar

Ship's radar scanner rotates.

Radar screen

Lighthouse

Land

Other ship

Radar is most often used to work out the positions of ships and aircraft. A beam of radio waves is sent out from a scanner, usually a rotating one. When the waves hit objects, especially metallic ones, some of their energy bounces back off the objects. From these "echoes" it is possible to plot a picture of the surrounding area and work out how far away things are. Radar is short for "radio detection and ranging".

Radio microphones

These are sometimes used at concerts or in TV studios to avoid having lots of trailing wires. The microphone has a transmitter inside it. Receivers are set up out of sight. They pick up the transmitted sound and feed it either to loudspeakers or to the control desk.

Bleepers

Some people, hospital doctors for instance, carry small radio receivers around in their pockets. When they hear a "bleep" on the radio, they know they have to go to, or should telephone, a certain place. This system is called radio-paging.

Model control

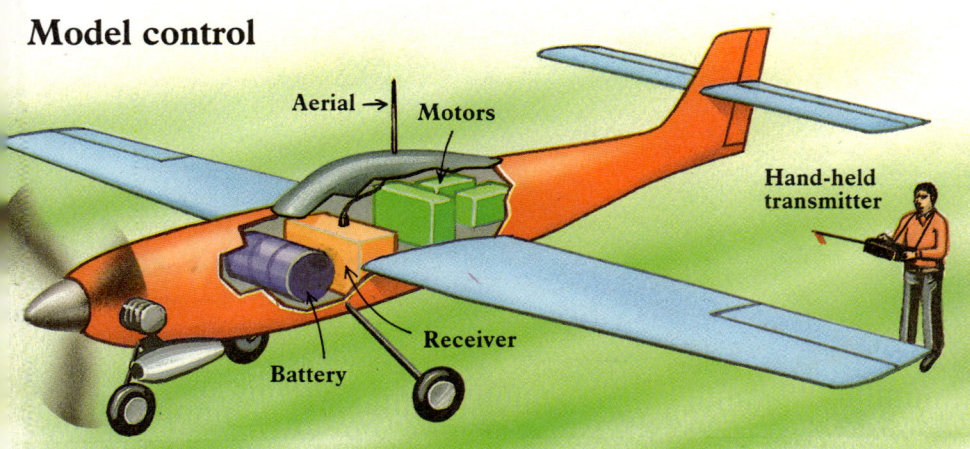

Model planes and boats can be controlled by radio. The operator uses a small transmitter to send out coded radio signals telling the model how to move. The model has a receiver in it. This interprets the coded instructions and passes them on to the motors which work the different parts of the model.

Sonar

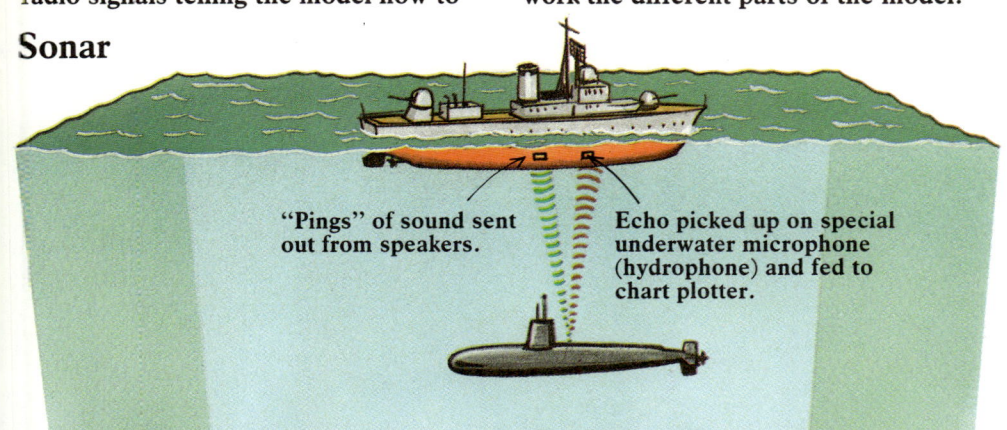

Most radio waves do not travel well through water but sound waves do. Ships can find out if there are submarines nearby by sending out bursts of sound. If the sound hits a submarine, an echo comes back. The position of the submarine and its distance away is then automatically plotted on a chart. The same method is used to find out how far down the sea-bed is, and to locate shoals of fish. Sonar is short for "sound navigation and ranging".

Weather balloons

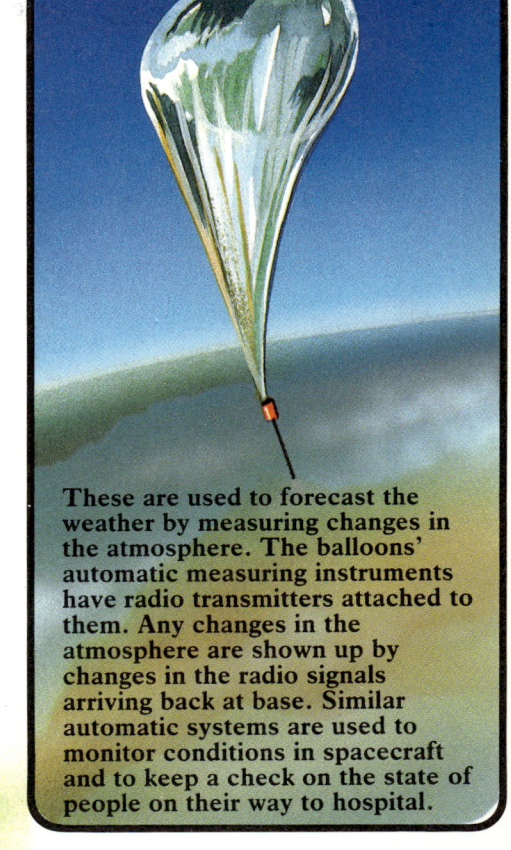

These are used to forecast the weather by measuring changes in the atmosphere. The balloons' automatic measuring instruments have radio transmitters attached to them. Any changes in the atmosphere are shown up by changes in the radio signals arriving back at base. Similar automatic systems are used to monitor conditions in spacecraft and to keep a check on the state of people on their way to hospital.

Aircraft landing

Radio is very important in helping planes to navigate. Coded radio signals are sent out automatically from airports. These tell pilots whether they are on course, first for the airport, then for the runway. One signal tells them whether they are right or left of the runway, another whether they are too high or too low.

93

The future

Experts are always trying to find new ways of improving the quality of sound recording and reproduction and to develop new techniques for radio broadcasting. Two of the most revolutionary advances are, in audio, the invention of the compact, digital disc and, in radio, the use of satellites. These are described on the right, along with some of the other things you can expect to find in a home of the future.

Turntables will play "compact discs". These exist already. They are only 120mm in diameter and run for an hour a side at a speed of several hundred r.p.m. The discs are recorded digitally and, instead of a groove, they have millions of microscopic "pits and flats". Turntables for the discs have a laser beam instead of a stylus. The beam scans the surface of the disc, "reading" the pattern of pits and flats. A digital current is produced, which is later converted to a normal current and then into sound.

Pit
Flat

▲ Cassette decks will include the digital process in playback and recording, and produce improved quality sound.

Some wrist watches will have radios in them.

▲ You will press a control on your self-tuning radio to indicate what type of programme you want. The tuner will then automatically scan through the different stations, playing you a few seconds of each appropriate one. When you hear a programme you like, you will press a hold button. If you also press the memory button, the programme will remain stored in the radio's computer memory, ready for re-call.

▲ All audio equipment will be remote controlled. Coded instructions will be sent on invisible infra-red rays from a handset to a special receiver in the equipment. This will decode the instructions and pass them to the appropriate controls.

Digital sound

Processing sound digitally (by numbers) cuts out distortion almost completely. The process can already be used in making tapes and records and will also be used for radio and TV sound. The electric signals from a microphone are scanned several thousand times a second and are given a number depending on their strength at each moment of scanning. The string of numbers, in the form of electric pulses, can travel by wires or radio, even over great distances, without the quality of the final sound suffering at all. This digital current has to be converted back into the normal type of current, called an analogue current, before being fed to loudspeakers.

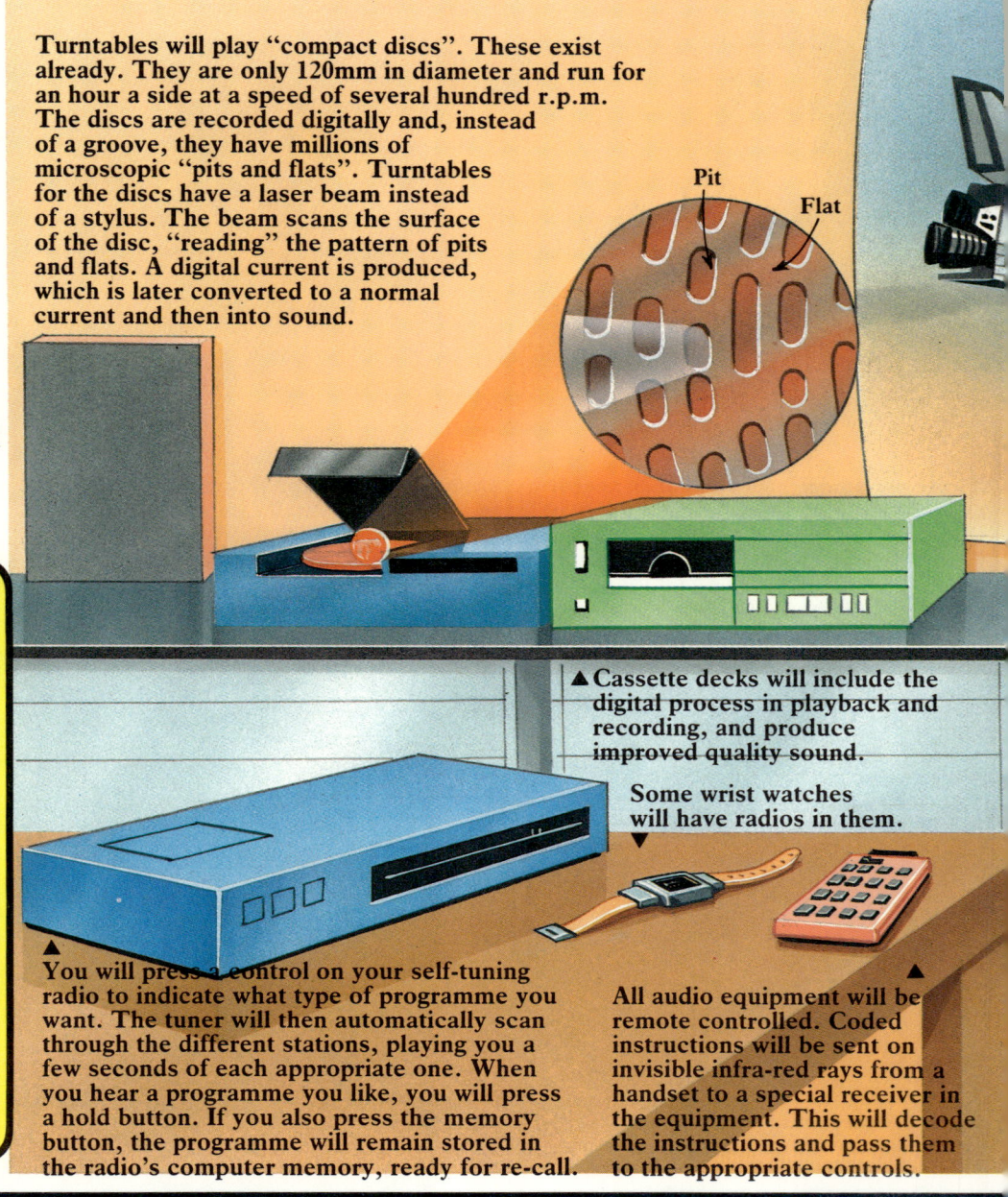

Audio and radio words

Here are some audio and radio words and their meanings. If the word you want is not shown in this list, look it up in the index, as it may be mentioned elsewhere in the book.

ANALOGUE. An electric current which flows smoothly, unlike a digital current, which is in pulses.
ANTENNA. Aerial.
AUTOREVERSE. Cassette deck which plays or records both sides of a tape without you having to turn the cassette over.

BANDWIDTH. The range of sound frequencies that a piece of equipment can handle. Measured in hertz. The wider the range the better. Don't confuse bandwidth with waveband.
BIAS. A special signal used in tape recording to avoid distortion. Some cassette decks have a switch for you to select the right bias level for the type of tape you are using, some do it automatically. Also, a control in turntable tonearms which helps the stylus sit centrally in the record groove.
CROSSOVER POINT. The particular frequencies at which signals to a loudspeaker are divided up to be handled by the different sorts of speaker (tweeters, squawkers, woofers).

DECIBEL (dB). A unit of measurement of the loudness of sounds, or the size of electric signals.
DIGITAL. An electric current which is in a series of pulses.
DIGITAL DISPLAY. A display given in numbers, as on the frequency display of tuners and receivers.
DIN. A German organization which lays down internationally recognized standards for audio equipment. Anything which meets the DIN standard 45500 qualifies as hi-fi.
DISTORTION. The difference between the signal which goes into an amplifier and the one that comes out. Measured as a percentage. Look for a "total harmonic distortion" (THD) figure which is as low as possible.
DUAL CAPSTAN. Cassette deck with two

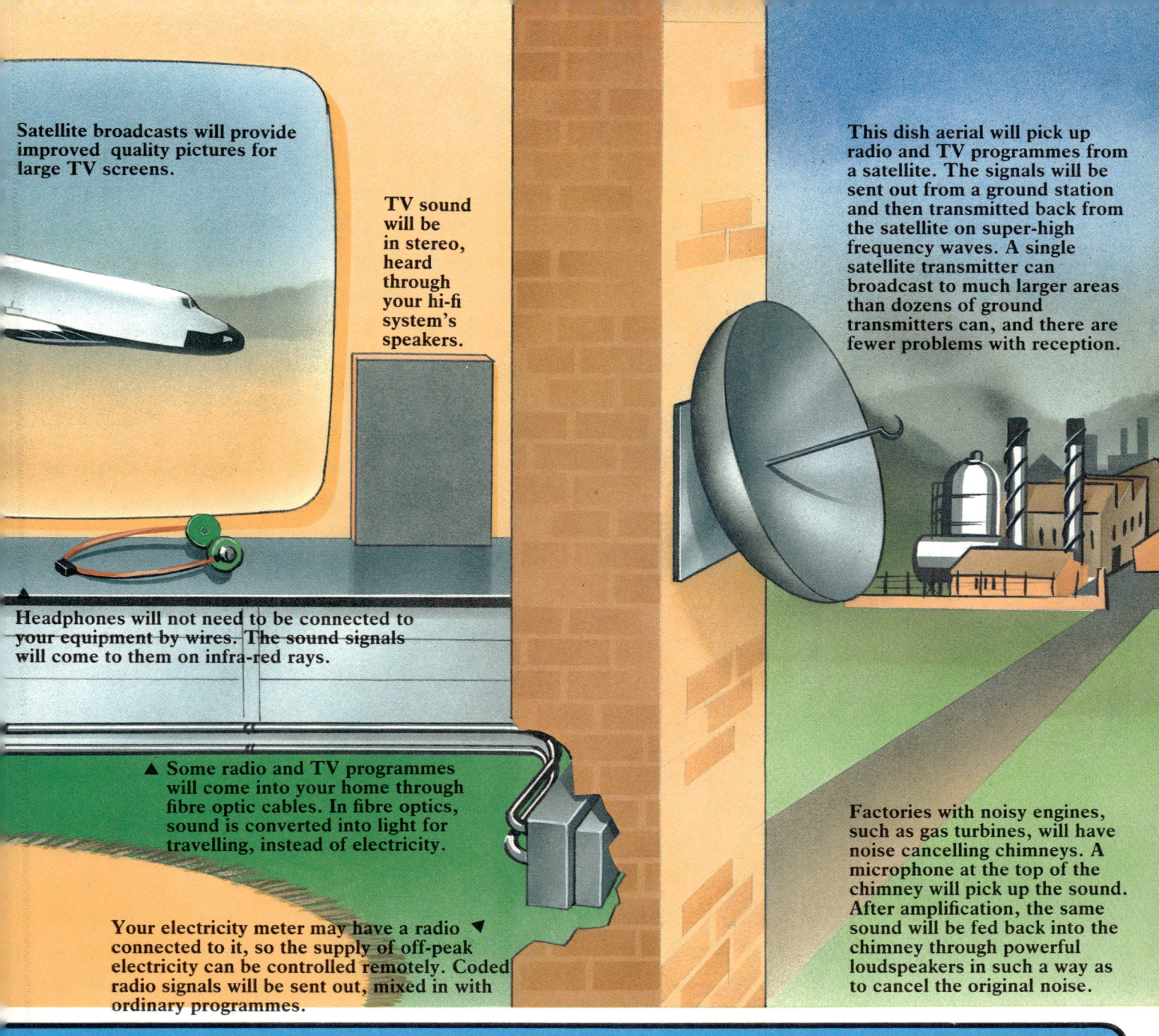

Satellite broadcasts will provide improved quality pictures for large TV screens.

TV sound will be in stereo, heard through your hi-fi system's speakers.

This dish aerial will pick up radio and TV programmes from a satellite. The signals will be sent out from a ground station and then transmitted back from the satellite on super-high frequency waves. A single satellite transmitter can broadcast to much larger areas than dozens of ground transmitters can, and there are fewer problems with reception.

Headphones will not need to be connected to your equipment by wires. The sound signals will come to them on infra-red rays.

▲ Some radio and TV programmes will come into your home through fibre optic cables. In fibre optics, sound is converted into light for travelling, instead of electricity.

Your electricity meter may have a radio ▼ connected to it, so the supply of off-peak electricity can be controlled remotely. Coded radio signals will be sent out, mixed in with ordinary programmes.

Factories with noisy engines, such as gas turbines, will have noise cancelling chimneys. A microphone at the top of the chimney will pick up the sound. After amplification, the same sound will be fed back into the chimney through powerful loudspeakers in such a way as to cancel the original noise.

capstans for pulling the tape past the heads. This helps keep the tape speed constant.
FREQUENCY RESPONSE. Same as bandwidth. See opposite page.
GIGAHERTZ (GHz). 1GHz = 1,000MHz.
HERTZ (Hz). The unit of measurement of frequency. 1,000Hz = 1kHz.
HF. High frequency. The high frequency waveband ranges from 3 to 30MHz.
IMPEDANCE. The resistance of certain components in a circuit to the flow of electricity. Measured in ohms. Loudspeakers and headphones must have the correct impedance for the system's amplifier to prevent distortion of sound or damage to the equipment.
KILOHERTZ (kHz). 1kHz = 1,000Hz.
LF. Low frequency. The LF waveband ranges from 30 to 300kHz.
LOW MASS. Lightweight tonearm and cartridge.
MEGAHERTZ (MHz). 1MHz = 1,000kHz.
METAL COMPATIBLE. Cassette deck which will play metal type tapes.
MF. Medium frequency. The MF waveband ranges from 300kHz to 3MHz.
MONO. One-track sound, that is with only one speaker or microphone.
PHONO. Label on amplifier switch which selects signals from turntable.
PICKUP ARM. Tonearm.
SENSITIVITY. The amount of signal which has to be fed into a piece of equipment in order to produce a given amount of power in the loudspeaker.
SEPARATION. The amount by which the signals from one channel of, say, a stereo system will "leak" on to the other, producing a slight mixture in the speakers. Measured in dB. The higher the figure the better.
SHF. Super-high frequency. SHF radio waves range from 3 to 30GHz.
SIGNAL TO NOISE RATIO. The size of the wanted signal compared with unwanted noise produced in the equipment's circuits. Measured in dB. The higher the figure the better.
STEREO. Two-track sound system, with two microphones and two loudspeakers.
UHF. Ultra-high frequency. UHF radio waves range from 300MHz to 3GHz.
VHF. Very high frequency. The VHF waveband ranges from 30 to 300MHz.
WAVEBAND. A particular range of radio frequencies, e.g. LF, VHF.

Index

aerial, 37, 46, 60, 67, 69, 71, 72, 74, 75, 76, 77, 83, 90, 91, 92, 94, 95
   dish, 92, 95
aircraft, 92, 93
"Altair" computer, 22, 30
amateur radio, 91
amplifiers, 46, 75, 82, 83, 85, 86, 87, 94, 95
   integrated, 85
   power, 85
amplitude, 72, 74, 75
   modulation, 74, 75
analogue
   current, 94
   signal, 56, 57
analogue-digital converter, 56
analytical engine, 2
AND gates, 9
architects, 22
arithmetic unit, 5, 7
artificial intelligence, 18
astronomy, 92
audio dub, 60
automatic edit, 60

Babbage, Charles, 2
backing store, 10, 11, 31
balance, 85, 87
bandwidth, 94, 95
bar codes, 3
barn doors, 42
BASIC programming language, 13, 15
bass, 87
battery, 72, 73, 76, 77, 82, 88, 90, 93
Betamax, 61
bias, 87, 94
binary code, 8, 10, 11, 13, 29, 31
bits, 10, 31
bleepers, 93
booms, 41, 43
brain, 2, 19
broadcasting, 36, 37, 46
bubble memories, 11
bugs, in programs, 12, 31
buses, 45
bytes, 10, 31

cable TV, 46, 47
calculators, 7, 30
camera tube, 63
cameras, 22, 35, 38, 39, 61, 62
   colour, 40
capacitors, 76, 88, 89, 90
capstan, 84, 85, 86, 94
caption rollers, 44
car audio, 83
cartoon films, 28
cartridges, 84, 87, 95
   moving coil, 84, 86
   moving magnet, 84, 86
cassette
   decks, 83, 84, 85, 86, 87, 94, 95
   recorders, 81, 83
cassettes, 11, 25, 61, 71, 83, 85, 87
   making of, 78-81, 91
central processing unit, 5, 7, 31
channels, 49, 61
channel selector, 61
character generators, 44, 57
chess, 2, 18
chips, 6-7, 8, 11, 22, 30
chromakey, 54, 55
chrome dioxide (CrO$_2$) tape, 85, 87
chrominance signals, 40

circuits, 6, 7, 8, 9
clock, 5, 7
COBOL programming language, 13
commentators, 50
compact discs, 94
computer
   crimes, 25
   games, 2, 7, 22, 23, 25, 27
   pictures, 9, 28-29
   programs, 59
   simulation, 26-27
computers, 34, 56, 57, 58, 59, 92, 94
   home, 32, 34, 58, 59
continuity announcer, 68
   desk, 68
control desks, 44, 45, 68, 70-71, 78-79, 80, 93
   rooms, 37, 42, 43, 44, 45, 68, 70-71, 78-79
counterbalance, 87
CPU, 5, 7, 31
cuts, 45
cutting head, 81
cutting room, 53

data, 2, 4, 5, 10, 11, 12, 16, 24, 30, 31
databank, 11, 24, 31
data storage cabinets, 4
decimal numbers, 8
delay-box, 71
demodulation, 77
diamond stylus, 81, 84
diaphragm, 41, 67
dichroic mirrors, 40
digital
   current, 94
   delay, 80
   display, 86, 87
   effects, 56, 57
   paint boxes, 57
   records, 85, 94
   signals, 56
   sound, 94
   watch, 21
digital-analogue converter, 57
digits, 8
diodes, 76, 77, 88, 89, 90
   detector, 77
   light emitting (LEDs), 87
Direct Broadcasting Satellites, 47
director, 42, 43, 44, 51
disaster relief, 17
disc jockey, 70-71
discs, video, 35, 61
display screens, 2, 3, 4, 23, 25, 29
distortion, 86, 87, 94, 95
doctors, 16, 22, 24
dolby, 87
dolly, 38
dots on TV screen, 48, 49
double shot, 54, 55
drawing robot, 28
dubbing, 53

ears, 2, 19
echo, 79, 80
edit start button, 53
editing, 52, 53, 70, 78, 79, 85
electromagnetic spectrum, 73
electron gun, 38, 39, 48
electronic
   components, 6, 7
   computers, 4, 6, 30
   effects, 54, 55
   field production, 51

games, 7, 22
keyboard, 22
news gathering (ENG), 50
viewfinder, 38
electrons, 39, 48
engineer, 68, 70-71, 79, 80
   sound, 41, 50
   vision, 50
ENIAC, 4, 30
EPs, 86
equalization, 87
erase head, 79, 84, 85
erasing, 79, 85
EXPLOR programming language, 13
eyes, 2, 19
   of robots, 13

fader, 70, 79
Ferranti, Mark I, 30
ferro-chrome tape, 87
ferro tape, 87
fibre optic cable, 47
fibre optics, 95
film, 36, 50
   editing, 53
filming on location, 51
first computers, 4, 6, 30
floor manager, 43
floppy discs, 11
flowcharts, 14, 15, 31
flutter, 86
FORTRAN programming language, 13
frame – advance, 60
   freeze-, 60
   libraries, 57
frequency, 46, 49, 67, 72, 74-75, 77, 79, 86, 87, 91, 95
   modulation, 74, 75
frog simulator, 26

gamma rays, 73
gates, 9
ghosting, 47
gigahertz (GHz) 95
graphic equalizers, 87
graphics pad, 3
graphs, 3, 21

hardware, computer, 12, 31
HDTV, 63
headphones, 66, 68, 71, 78, 79, 82, 83, 84, 86, 87, 95
helical system, 52
hertz (Hz), 94, 95
Hertz, Heinrich, 73, 74
hi-fi, 66, 82, 94, 95
   mini, 82, 83
holography, 63
home video
   cameras, 38, 62
   equipment, 60, 61, 62
   tape editing, 53
houses, computerized, 22
hydrophone, 93

impedance, 84, 86, 87, 88, 95
inbetweening, 28
infra red, 73, 94, 95
   remote controls, 49, 61
input, 2, 3, 5, 7, 86
   types of, 3, 4, 23, 31
instructions to computer, 4, 5, 10, 12, 30

integrated circuits, 6-7, 30, 88, 89, 90
intelligence, of computer, 19
interference, 73, 75
interviews, 70, 71
ionosphere, 74, 75
iron oxide, 78, 79

jackfield, 71
jingles, 68, 70, 71

keyboards, 2, 3, 4, 15, 17, 23, 29
kilohertz (KHz), 74-75, 95

land line, 50
lasers
   beams, 3, 11, 19, 94
   discs, 11
lenses, 39, 40
libraries, 22, 23
light, 73, 92, 95
linear tracking, 84
lines, on TV screen, 39, 48, 49, 63
LISP programming language, 13
live broadcasts, 36, 44
loader, 60
long-term memory, 10
loudness control, 87
loudspeakers, 66, 68, 69, 71, 76, 77, 79, 80, 83, 84, 85, 88, 90, 94, 95
LPs, 86
luminance signal, 40

magnet, 67, 78, 81, 84, 85
magnetism, 67, 79
mainframes, 4, 24, 30
Manchester University Mark I, 4, 30
MCR, 50
megahertz (MHz), 75, 95
memory, 86, 94
   chips, 7
   human, 10, 19
   store, 5, 7, 10-11, 13, 19, 29, 31
metal tape, 85, 87, 95
microchips, 6-7, 8, 11, 22, 30, 34, 35, 56, 58, 63
microcomputers, 4, 11, 23, 32
Micromouse, 18
microphones, 3, 18, 23, 37, 41, 42, 43, 44, 51, 67, 70, 71, 75, 78, 79, 83, 87, 93, 94, 95
   bidirectional, 41
   capacitor, 85
   cardioid, 41, 85
   condenser, 85
   dynamic, 85
   gun, 50, 51
   moving coil, 41, 85
   omnidirectional, 41
microprocessor chips, 7, 16, 17, 18, 22, 31
microwave radio link, 51
minicomputers, 4
mixer, 77, 78-79, 80
mixes, 45
mixing desk, 55
mixing down, 80
mobile control room, 50
model control, 93
model train control, 16
modem, 59
modulation, 74-75
modulators, 46, 75
modules, 79, 80

monitors, 42, 44, 50, 58
mono, 82, 83, 87, 95
moving heads, 52
music, 17, 28
    centres, 82

navigation, 92, 93
networks, 47
news bulletins, 70
    gathering, 50
    programmes, 44
noise
    cancellation, 95
    reduction, 87
NOR gates, 9
NTSC, 40

ohms, 88, 95
optical viewfinder, 38
OR gates, 9
oscillators, 46, 75, 77
output, 2, 3, 5, 7, 21
    types of, 3, 4, 31
outside broadcasts, 50, 51
overdubbing, 79

paintings, by computer, 29
"painting by numbers", 55
PAL, 40
pan pot, 79, 80
panning, 62
peak indicators, 87
    limiter, 87
persistence of vision, 49
phase-locked (PLL), 86
phone-ins, 71
phosphor, 48
picture tube, 48, 63
PILOT programming language, 13
pinch wheel, 84
pitch, 67, 84
pixels, 29
playback head, 84, 85
playing back, 85, 86, 87
plotters, 3, 4
pocket computers, 23, 30
POP 2 programming language, 13
power indicators, 86
    output, 86
    supply, 5, 6, 7, 19
preamplifier, 85
prerecording, 36
printed circuit boards, 7
printers, 3, 4, 23, 31
producer, 70, 78-79
production control
    area, 50
    room, 42, 44, 45
production secretary, 70, 71
programmable car, 17
programmes, (TV & Radio), 68, 70-71, 83, 87, 94, 95
programming, (computers), 14, 23
programming languages, 13, 15, 31
programs, 12-13, 18, 19, 20, 23, 24, 29, 30, 31, 59
pulses, 4, 6, 8-9
punched cards and tapes, 11

'QSL' card, 91
quartz, crystal, 75
    lock, 86
    synthesizer, 86

rack systems, 82, 86-87
radar devices, 19
radio
    alarm clocks, 83
    cars, 71
    /cassette recorders, 82
    'ham', 91

paging, 93
societies, 91
station, 68, 74, 86, 88, 90, 94
studio, 68, 70-71
two-way, 71, 91
radios, 66, 69, 76-77, 82, 83, 88-89, 94, 95
RAM, 29, 31
Random Access Memory, 10, 31
Read Only Memory, 10, 31
receiver, 67, 76, 83, 91, 92, 93, 94
reception, 75, 88, 90, 95
recording, 52, 78-79, 83, 86, 87, 94
    heads, 52, 60, 78, 84
    levels, 87
    studios, 78-79
record press, 81
records, 70, 71, 85, 94
    making of, 78-81
reel-to-reel recorders, 68, 70, 78, 79, 80, 85, 86
relay station, 47
    transmitters, 69
remote controls, 49, 61, 94
    studios, 71
reporter, 70, 71
resistors, 76, 88, 89, 90
robots, 13, 18, 19, 20, 28, 31
roller caption, 57
ROM, 10, 12, 31
r.p.m., 86, 94
rumble, 84, 87

sapphire stylus, 84
satellites, 9, 19, 24
    broadcasting, 75, 94, 95
Saturn, 9
scratch, 87
screen, 48
screen pictures, 29
SECAM, 40
senses, 2, 19
    of robots, 19
Senster, 19
sewing machines, 22
shadow mask tube, 48
"Sharp PC1211", pocket computer, 30
SHF, 75, 95
ships, 91, 92, 93
ship's bridge simulator, 26
short-term memory, 10
shots, 53, 62
signal strength indicator, 86
silicon, 6, 30
software, 12, 31
soldering, 88
soloing, 79
sonar, 93
sound
    control room, 42
    effects, 42, 71, 80
    engineers, 41, 50
    signals, 67, 68, 69, 74, 75, 76, 77, 78, 79, 80, 85, 86, 87, 92, 94
    synthesizers, 3, 17, 23
    waves, 66, 73, 85, 93
source, 85, 86
space, 72, 75, 92
spacecraft, 93
space pictures, 9
speed, 85, 86, 94, 95
speed of computer, 4, 5, 16
spools, 78, 79, 86
squawkers, 84, 94
stage manager, 42
stampers, 81
stars, 72
stereo, 80, 81, 82, 83, 85, 87, 95
    sound, 63
Stonehenge, 2

stroboscopes, 86
studios, 42, 43, 50
stylus, 81, 84, 87, 94
subsonic filter, 87
surround-sound, 83

tachers, 22
talking to a computer, 3, 18
tape (audio), 78-79, 84, 85, 87, 92
    counter, 86
    hiss, 87
    master, 80, 81
    recorders, 68, 70, 78, 79, 80, 85, 86
    types of, 85, 87
tapes (audio), 70, 80, 85, 94
target plates, 38, 39, 40, 56
telecine machine, 36, 37, 50
telephone cables, 24, 25
telephones, 22, 25, 67, 68, 71
    car, 91
teleprompt, 44
telescopes, 92
teletext, 25, 59
television, 73, 85, 92, 93, 94, 95
    films on, 49
    flat screen, 63
    future, 59, 63
    games, 58
    interactive, 59
    pay systems, 47
    satellite, 46, 47, 73
    sets, 48
    stations, 43
    three dimensional, 63
    tube set, 48
television sets, as computer
    terminals, 24, 25, 90
time-signals, 68
tone, 68, 71, 79, 80, 85, 87
tonearms, 84, 87, 94, 95
track, 78, 79, 80, 81, 85
tracking force, 87
transceiver, 91
transformers, 76, 88, 89
transistors, 6, 8, 9, 30, 76, 85, 88, 89, 90
translation, 17
transmitters, 37, 47, 51, 67, 68, 69, 71, 72, 74, 75, 76, 77, 88, 90, 91, 93, 95
transmitting, 37, 46, 69, 74-75
treble, 87
tube, camera, 38, 39, 40
tuner, 49, 60, 77, 82, 83, 86, 88, 90, 94
tuning, 69, 74, 77, 86, 94
turntables, 71, 81, 82, 83, 84, 86, 87, 94, 95
tweeters, 84, 94

UHF, 46, 74, 91, 92, 95
ultra-violet, 73
use of computers
    for the handicapped, 16
    for translation, 17
    in hospitals, 16
    in schools, 13, 16, 23, 24

valves, 6
VCRs, 52, 53, 60, 61
    formats of, 61
    portable, 61
VDU, 58
Veroboard, 88, 89, 90
VHF, 46, 69, 74, 76, 77, 83, 86, 87, 91, 92, 95
VHS, 61
video
    camera, 61
    discs, 35, 61

games, 34, 58
tape, 35, 52, 61
tape editing, 53
Video 2000, 61
videotex, 25
videotext, 59
view data, 59
vision
    control room, 43
    engineer, 50
    mixer, 43, 44, 45
visual display units (VDUs), 3, 4, 31, 90
voice-overs, 70
volume, 68, 70, 71, 76, 77, 79, 87, 88, 90
VTRs, 52
VU meters, 68, 79, 87

washing machines, 7, 22
waveband, 72, 74-75, 76, 77, 82, 83, 86, 87, 91, 92, 94, 95
wavelength, 46, 72, 73, 74-75
waves
    carrier, 48, 49, 69, 72, 74, 75, 76, 77
    electromagnetic, 72
    long, 69, 72, 74, 75, 76, 77, 83, 87
    medium, 69, 72, 74, 75, 76, 77, 83, 87, 88
    short, 74, 91
    sound, 66, 73, 85, 93
weather balloons, 93
    forecasting, 16
wipes, 45
woofers, 84, 94
wow, 86

zoom lens, 38, 62

Printed and Bound in Great Britain by Purnell and Sons (Book Production) Ltd., Paulton, Bristol

The name Usborne and the device are Trade Marks of Usborne Publishing Ltd.

First published in 1982 by Usborne Publishing Ltd, 20 Garrick Street, London WC2 9BJ, England.
© 1982 Usborne Publishing

All rights reserved. No part of this publication may be reproduced, stored in a retrieval system or transmitted in any form or by any means, electronic, mechanical, photocopying, recording, or otherwise, without the prior permission of the publisher.